You and Your Telephone

by
Tom Rogers

Howard W. Sams & Co., Inc.
4300 WEST 62ND ST. INDIANAPOLIS, INDIANA 46268 USA

FIRST EDITION
FIRST PRINTING—1980

International Standard Book Number: 0-672-21744-9
Library of Congress Catalog Card Number: 80-52229

Printed in the United States of America.

Preface

This book was not written to tell you how to "beat" your local telephone company. Rather, it was written to explain how your telephone works, what features and services you can expect to have in your home, and, utilizing this basic knowledge, how you can make the best decisions for your own circumstances. With this intent in mind, I have been very careful to avoid specific product endorsements. Why?

For many years, I have been involved in the subscriber side of the telephone industry. Although service offerings from some telephone companies may have seemed very limited in the past, this is not true of the present and definitely will not be true in the future. Decisions made by the federal regulatory agency in charge of telecommunications—the Federal Communications Commission (FCC)—now allow you to own your own telephone equipment. These decisions have opened up the market to many independent manufacturers and suppliers of telephone equipment. What this comes down to is that you can either lease or own your own telephone. You can also own many types of telephone accessories. Literally hundreds of these devices have appeared for sale in the last few years, and more continue to appear. By the time you read this, specific items may have changed. I do not want you to purchase obsolete old-style devices or unnecessarily expensive devices. Hence, I will describe many things you can do with your telephone, and accessory devices you can obtain without reference to specific manufacturers. However, for your convenience,

I have included an up-to-date listing of telephone equipment manufacturers and suppliers in the back of this book. You should be able to easily find retailers offering these devices. If not, contact the manufacturer for his recommendation of the nearest retailers. Very probably your local telephone company will sell them to you. Compare what your local company offers with devices you can buy at many local radio supply stores, almost all large department stores, many specialty shops, etc. The choice is yours!

What can you own? How can you tell if the device you want is FCC approved? What is the "registration statement" that must be on devices you purchase?

The telephone company is a public utility and has to furnish you at least basic telephone service. More than that, and contrary to what you may occasionally hear, the telephone company is *your* telephone company and will assist you to the best of its ability. You have to decide if that ability meets your needs. Can you buy the same service cheaper from another source? Will the end result be what you want? This book will tell you how your telephone works and what you can do with it. My purpose is to inform you so that you can better decide what you want to do. Whether you simply want to be the "first on the block," or really need the telephone accessories—you have to decide.

I welcome you to the world of useful and exciting telephone services!

Tom Rogers

To my family, and especially to my wife

Contents

1

Should You Buy or Rent?

Why do you want to own your own telephone? Can you install your own telephone as well as the local telephone company can? These and other questions must be running through your mind.

Yes, it is legal now to own your own telephone set and various other equipment. By purchasing, instead of renting as you do now, you may be able to save quite a bit of money during the life of the equipment. Also, it is possible to purchase more technically advanced, feature-packed equipment than the equipment that can be rented from the local telephone company. This situation may change or at least the technological gap may narrow as the telephone company becomes more competitive. As with any business, telephone companies respond to customer desires. However, they are usually slower than private retailers to offer innovations.

Many features are available on a wide variety of sets (Fig. 1-1). These telephones offer push-button dialing even though your local telephone service is not equipped for it; abbreviated dialers that allow you to dial stored telephone numbers by dialing only one or two digits; last number recall; built-in calculators, clocks, and timers; call-hold devices; speakerphones; portable radiotelephones; call accountants that tell you how much your long-distance call is costing while you are talking; and numerous decorator styles and features limited only by your imagination and ingenuity.

If you purchase quality, FCC-approved equipment from a reputable supplier, your telephone set, or whatever special devices you purchase for use with your telephone set, will be as reliable as your local telephone company equipment. In fact, if you are served by what is termed an *independent* telephone company—a non–Bell-System company (there are approximately 1700 independents in the United States)—the telephone company may be

Fig. 1-1. An assortment of modern telephone sets.

leasing you the same equipment that you can buy. Now that telephone companies have opened retail stores, you can purchase a telephone set from the telephone company! The catch for Bell-System customers is that Bell will only sell you the housing—they still rent you the inner workings! The housing is yours forever, but you must let them remove their parts if you move out of their service area. It is reasonable to say that any telephone equipment purchased from either your local telephone company or a reputable supplier will be good, reliable equipment. Telephone companies do not normally sell at a discount, so you may be able to purchase similar equipment from a reputable supplier at a substantial savings. Just be sure you receive a warranty with your purchase, and that your supplier has a good reputation for servicing your equipment should the need arise.

Watch for sales (Fig. 1-2). Special prices are run on overproduced equipment (the manufacturer or supplier has more on hand than he wants) and you may be able to save quite a bit due to these "inventory reduction" sales. Remember that a telephone is not a mysterious device any more than a radio is. People shop for radios—shop for your telephone!

You can install your own telephone set. The telephone companies are making this easier all the time. The newest wiring installations have small minimodular receptacles (Fig. 1-3). It is very inexpensive to have new houses prewired with receptacles (jacks) installed in every room. You will find that extra receptacles are inexpensive during the initial installation but relatively expensive later on. Plan ahead if possible! Extra unused receptacles are not gaudy and can be ignored like a regular wall electrical outlet.

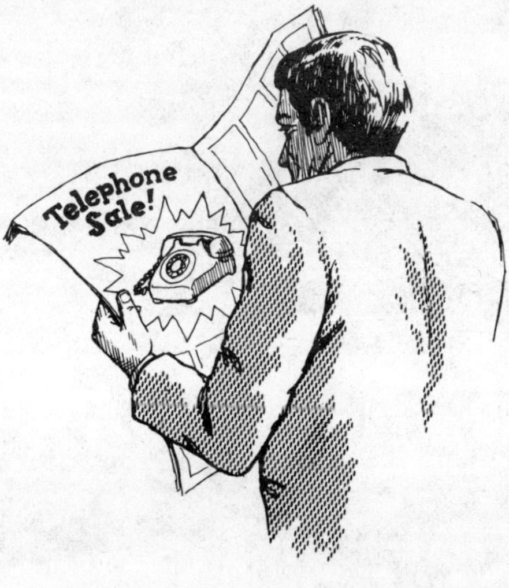

Fig. 1-2. Special sales can provide savings.

All prewiring can be easily installed inside your walls, whereas later installation may result in "surface" wiring (wiring run on the visible side of a wall). Surface wiring may not be objectionable; it usually blends into the room decor especially when painted to match the room. But, if possible, it will be to your advantage to have your house fully prewired initially.

You are not normally allowed to install the wiring inside your house. You should check with your local telephone company to determine if they will permit you to do this wiring. In most cases you would be adding to existing telephone company house wiring. If they allow you to do this wiring, please read Chapter 9—*Home Wiring*. Never modify, move, or alter in any way the small box—usually gray or black—located where the outside wiring from the telephone company enters your house (Fig. 1-4). This is a "station protector" and includes devices to protect *you and your home* from possible hazardous voltage from lightning during a storm. Do *not* remove the ground wire—usually a single, relatively

(A) Surface mount receptacle.

(B) Flush mount receptacle.

Fig. 1-3. Minimodular receptacles.

Fig. 1-4. Station protector installation.

large wire (No. 12 AWG) that is either insulated or bare copper—running between this box and an electrical ground in your house! Removal of this wire will usually nullify your lightning protection! Remember, the station protector is there for *your* protection—treat it with kindness.

Telephone service is available in a number of ways. Some of these services may not be available in certain geographic regions due to limitations imposed on the local telephone company.

Most readers will have single-party residential service. Multiparty or "party-line" service exists with up to 10 homes sharing the same telephone line. The major telephone companies have programs underway to reduce or eliminate multiparty service, but we have to consider it. I will point out differences in installation and equipment whenever they arise. Currently only single-party service may have customer-provided equipment—check with your local telephone company if you have multiparty service to determine what it allows.

We are now ready to learn how your telephone works. We will not discuss complicated details, but we will present a basic understanding of the telephone system to enable you to select better telephone equipment that will provide you with the features and services you desire.

Note that when you purchase items for your telephone system you will get a detailed set of installation instructions. Keep these instructions for future use.

2

The Telephone System

Your telephone can connect you to any other telephone (Fig. 2-1). This is really amazing when you think about it!

When you lift the telephone handset, a switch inside the telephone "turns on" your telephone. (The ringer of your telephone remains connected to the telephone line all the time; that is why your telephone can ring when the rest of the set is "turned off.") Now that your telephone is "on," the wires leading from it eventually go to a telephone building known as a *central office* (Fig. 2-2). The central office is interconnected to other central offices much like a spider web (Fig. 2-3). The "End Office" is your serving central office, where your own telephone number is assigned.

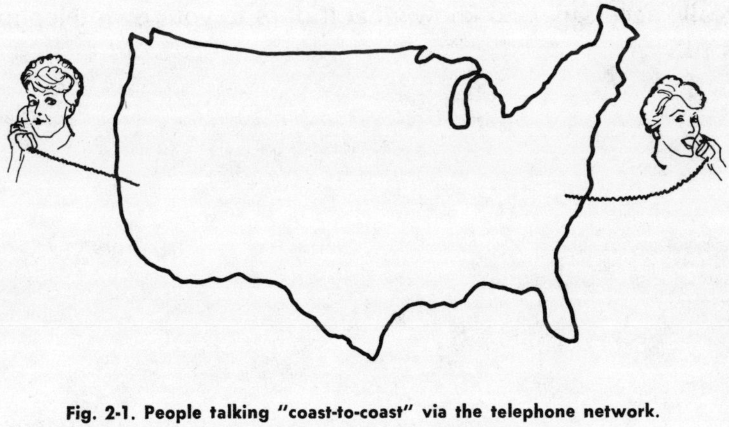

Fig. 2-1. People talking "coast-to-coast" via the telephone network.

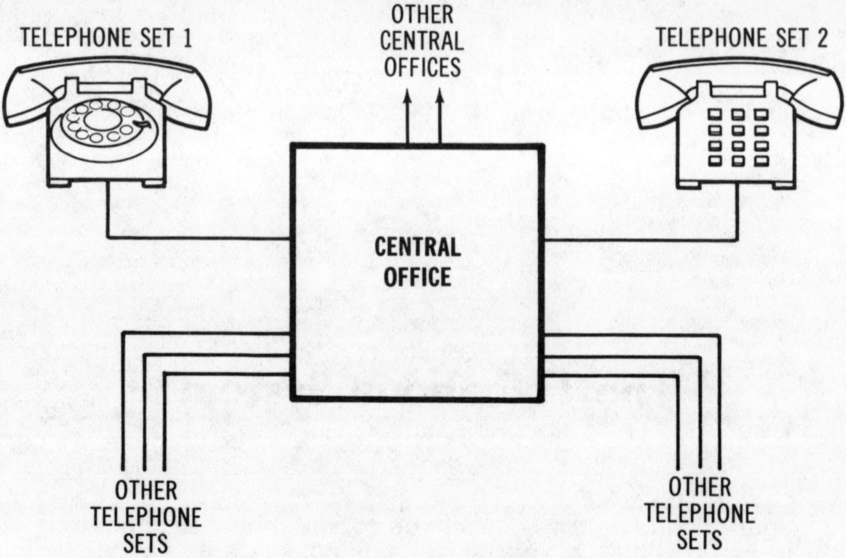

Fig. 2-2. Telephone system.

At each junction of two or more threads of the "spider web" lies a central office. For all practical purposes, the central office can be thought of as the brain—or one of them—of the telephone system. (As a public service your local telephone company probably provides guided tours of some of their central offices. This is an interesting tour that gives you a better insight into the workings of the telephone system.) An explanation of the interworkings of the central office will not be given here. Rather, all that is really necessary is to know what it does to your own telephone set.

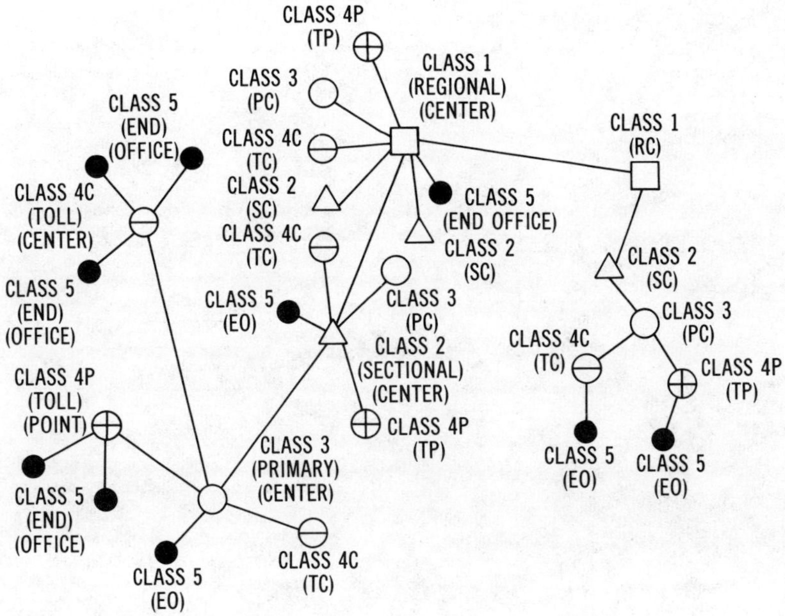

SYMBOL	CLASS	NAME	ABBREVIATION
□	1	REGIONAL CENTER	RC
△	2	SECTIONAL CENTER	SC
○	3	PRIMARY CENTER	PC
⊖	4C	TOLL CENTER	TC
⊕	4P	TOLL POINT	TP
●	5	END OFFICE	EO

Fig. 2-3. Telephone network.

3

Old Telephones

Who can resist the romance and nostalgia associated with a magneto telephone (Fig. 3-1)? Or what about the three-slot coin telephone? Of course there are numerous "200," "300," and other old-style telephones still available. Should you use them? Will they work with today's modern telephone system? Let us see.

MAGNETO TELEPHONES

Many different housings were made to hold basically two versions of this antique. The oldest, called *local battery,* that you are likely to find had a battery compartment. The batteries were big dry cells, like large flashlight batteries, or else rechargeable wet cells, like small car batteries. If you can find two or more of these sets you can wire them together as an intercom. The drawing in Fig. 3-2 shows you how. Note that there is no provision made to signal (ring) either end. Current military field telephones use this basic circuit. The lack of a dial is but one of many obstacles that prohibit using this set with a modern telephone switching system.

The newer version is called a *common-battery* magneto telephone. It did not have a local battery compartment. Battery power was provided by the serving central office. This was seen as a major modernization at the time. (Imagine all those batteries that had to be constantly checked and replaced in a local-battery system!)

Two or more common-battery magneto telephones can be wired together to make an intercom. The drawing in Fig. 3-3 shows you how. Note that there is no provision made to signal (ring) either end. The common-battery system is still used today, but with much more modern telephone sets.

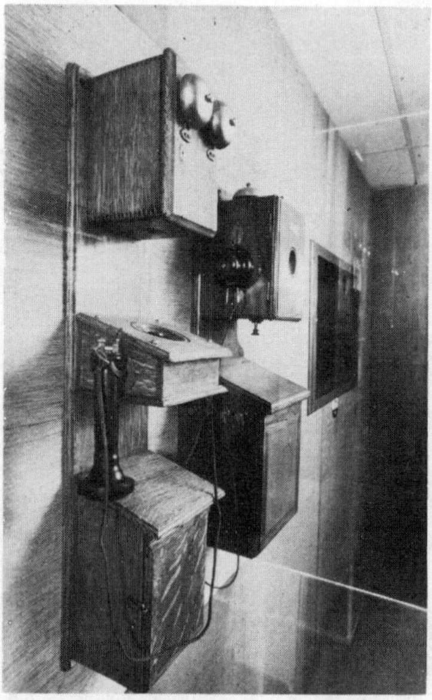

Fig. 3-1. Magneto telephone sets.

During these periods in our history there were no central offices containing sophisticated switching machines. All interconnection of telephone lines was done by human operators. "Hello Central" was a common greeting when you contacted your operator. Why was this necessary?

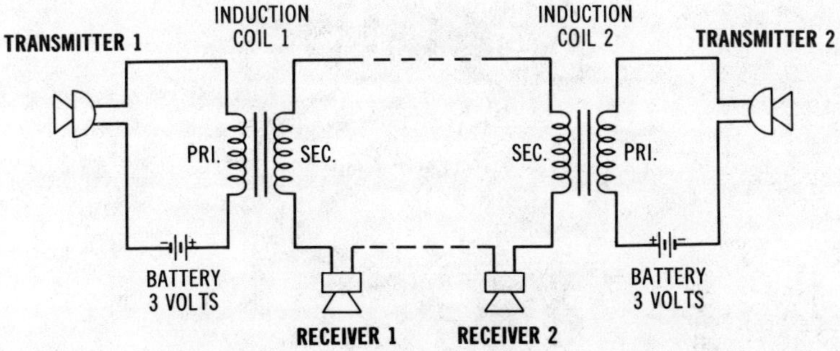

Fig. 3-2. Local-battery telephone system.

Magneto telephones do not contain any dial. (There were some magneto telephones manufactured with dials, but these were exceptions, not the rule.) They work on principles alien to our present telephone system. As attractive as they may seem to you, they will not work on your present telephone line.

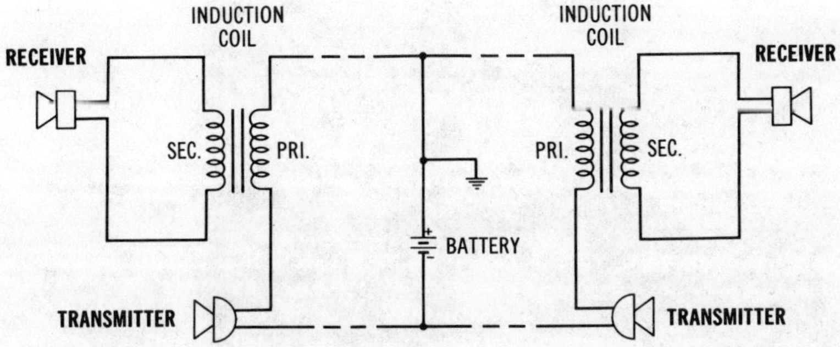

Fig. 3-3. Common-battery telephone system.

To take advantage of their appeal, many manufacturers offer magneto-style telephones with modern components hidden inside (Fig. 3-4). They are decorative and perform well.

As an aside, I want to mention a practice I have seen all too often. People have a magneto telephone passed down from their

(A) Cover closed.

(B) Cover open.

Fig. 3-4. Replica magneto telephone set.

grandparents or purchased cheaply a few years ago. By having standard telephone components installed, they can have the best of both worlds with a low cash outlay. Please *don't* do it. Magneto telephones are true antiques and are virtually ruined by any modification. Keep your antique. Purchase a replica with modern components. If you don't believe this makes good sense, call an antique dealer and find out how much your telephone is worth. You may be happily surprised. You can probably buy several replicas for the price of your genuine antique.

TOMORROW'S ANTIQUES

The Bell Telephone System's model designations of "200," "300," "candlestick," and others describe a type of telephone set that was very modern in its time (Fig. 3-5). Several of the major

Fig. 3-5. Old-style telephone set.

independent telephone manufacturers made sets similar to them. But no one makes originals anymore. They are becoming very scarce. Why is this happening when millions of them once existed?

Due to federal regulatory and telephone company policies, when a telephone set is removed from service it is either refurbished to modern specifications or junked. No one would want defective or obsolete telephone sets sold to a retailer who might sell them to people who would try to reuse them. So policies have been strict.

The "300" set obsoleted the "200" set. The modern day "500" set obsoleted the "300" set. It costs a lot of money to modernize or refurbish an old set. It may be cheaper to buy a new type. Therefore, due to these cost considerations, most older sets have been junked. And before they were junked they were destroyed, often by using a sledge hammer. Candlestick, magneto, etc., all had the same fate. Preserve yours.

4

Dialing

The telephone system uses two types of dialing—rotary dialing and dual-tone multifrequency (DTMF) dialing. They both perform the same function—telling your central office the number you are calling.

ROTARY DIALING

Rotary dialing is allowed everywhere, even if you have DTMF dialing. This is the common mechanical fingerwheel dial, with various speeds of operation used through the years (Fig. 4-1).

There are millions of these dialing mechanisms in use today. Very often the only way telephone sets are available from telephone refurbishing/rebuilding companies is with rotary dials. Rebuilt telephones have to pass the same specifications and tests that a new set does. With a good warranty they make the least expensive telephone sets available, and quite often can be purchased from local retailers.

Two photographs of a rotary dial telephone with various external parts removed, and with the cover removed and the handset disassembled, are shown in Fig. 4-2. You can unplug the cords, but you may not want to disassemble your set further. If curiosity gets the best of you, look at the photographs and leave your telephone alone.

How does a rotary dial work? When the fingerwheel is rotated and then released to allow it to return, a series of electrical pulses

Fig. 4-1. Rotary dial telephone.

is generated (these pulses are much like turning a light switch alternately off and on); one pulse for "1," two pulses for "2," and so forth to 10 pulses for "0." When you dial "0" for the operator you actually generate 10 pulses. These pulses are precisely controlled in duration and relationship to one another by the rotary dial mechanism. Since we are not going to be designing dials in this book, I am not going to trouble you with all of the numerous specifications for these dials. However, a somewhat more technical explanation is given in the following section.

ROTARY DIALING—TECHNICAL EXPLANATION

The rotary dial (Fig. 4-3) has a more complex job than most people realize. When you dial you do not want to hear the dial pulses. To prevent this, a set of contacts open and block the outgoing dial pulses from your receiver when the fingerwheel is rotated.

Similarly, to eliminate the internal resistance of the transmitter, another set of contacts close and shunt the transmitter. After dialing is completed, the contacts (called *off-normal* contacts) restore the circuit to "normal." The diagram in Fig. 4-4 illustrates this.

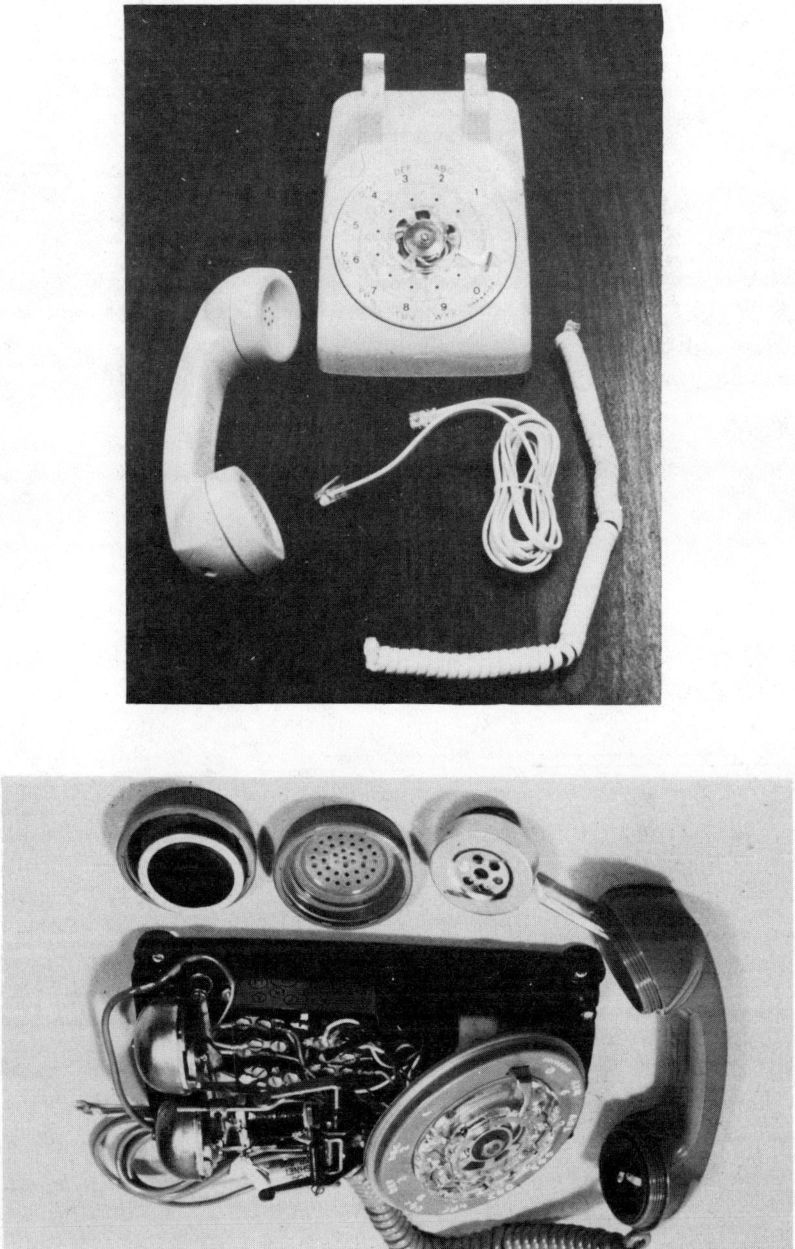

Fig. 4-2. Disassembled views of rotary dial telephone.

Fig. 4-3. Rotary dial mechanism.

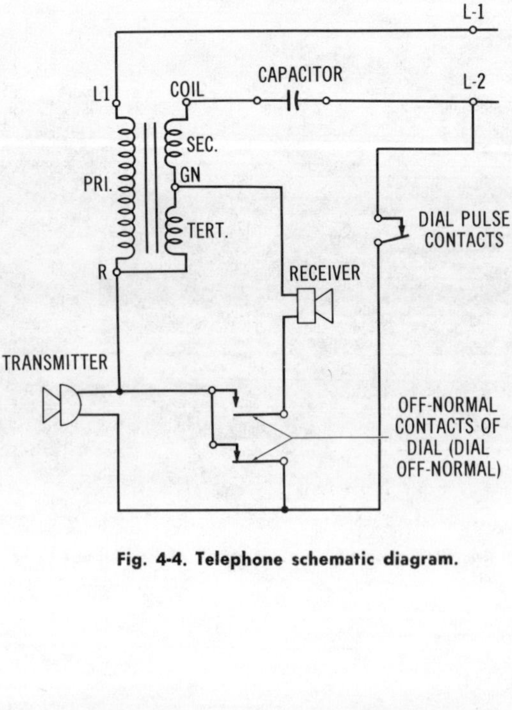

Fig. 4-4. Telephone schematic diagram.

We all know that as we rotate the fingerwheel, a spring is wound up that will return the fingerwheel to its rest position after we remove our finger. There is also a governor mechanism that controls the rate at which the spring returns the fingerwheel. The governor is nominally adjusted to allow return rotation at a rate that will generate 10 dial pulses per second from the "pulse" contacts of the dial. If you dial a "4," four pulses will be generated at a rate of 10 pulses per second.

The pulses are composed of "open" and "closed" time periods (called "break" and "make," respectively) as shown in Fig. 4-5.

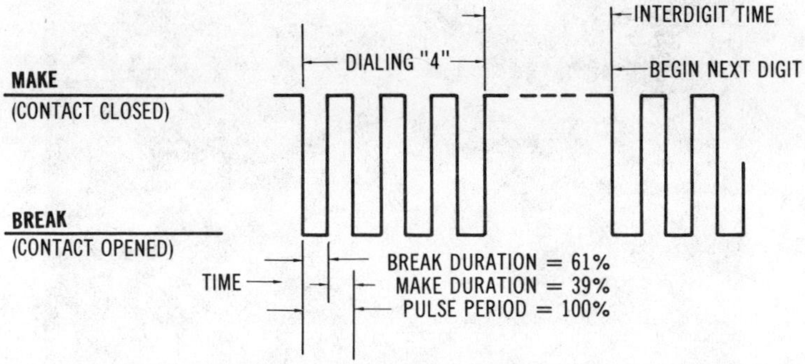

Fig. 4-5. Dial pulse duration.

Note that nominally 61% of the total pulse is composed of "break" and 39% is composed of "make." This is called the *percent-break* ratio. The percent-break ratio is deliberately not 50% in order to compensate for the differing electrical characteristics of transmission systems and central office switching equipment, and in order to make the best use of circuit conditions occurring during the break and make time intervals.

What about the time between digits when the dial is in the normal or rest position? Interdigit time is the time from the end of the last pulse of a dialed digit to the beginning of the first pulse of the succeeding dialed digit. Interdigit time varies from switching system to switching system, with 500 to 700 milliseconds being nominal. During this time the central office switching system recognizes the end of one dialed digit and prepares itself to receive the next.

REMOVING THE FINGERWHEEL

To clean or replace the fingerwheel of the rotary dial, rotate the fingerwheel as far clockwise as it will go and insert the end of a piece of stiff wire (such as a straightened paper clip) into the small hole located between and just above the "9" and "0" finger holes (Fig. 4-6). Press the wire into the hole until you release the tab at the bottom of the hole. You should feel the tab release. Further rotate the fingerwheel clockwise until it is released.

To replace the fingerwheel, position the "0" hole in the "9" at rest position. Push the fingerwheel gently down onto the dial.

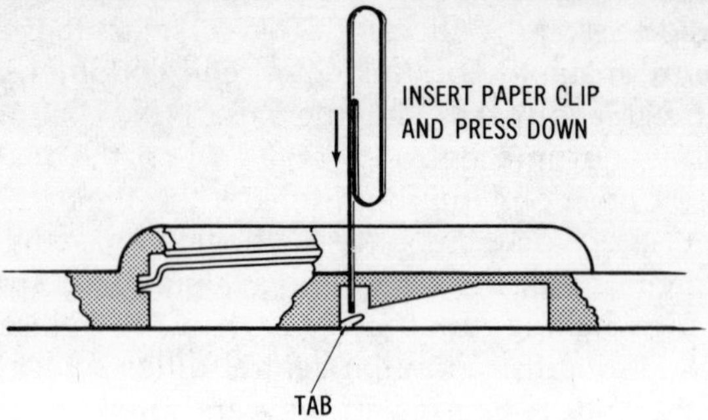

Fig. 4-6. Fingerwheel removal.

Rotate the fingerwheel in a counterclockwise direction until it locks into position. You should feel the tab snap into place to lock the fingerwheel onto the dial.

Fig. 4-7. Dual-tone multifrequency (DTMF) telephone.

DUAL-TONE MULTIFREQUENCY DIALING (DTMF)

DTMF dial-equipped telephones are the familiar Bell System Touch-Tone* telephones (Fig. 4-7). Other manufacturers use the same output specifications, but call theirs by other names. They all do the same thing. By depressing one button, a pair of tones (dual tones) of different frequencies is generated. The electrical combination of tones is sent to the central office where the end result is the same as the rotary dial. DTMF is faster to use. You can dial just about as fast as you can press the buttons; allowing approximately 1/2 second for interdigit time as explained under "Rotary Dialing—Technical Explanation."

Additionally, two nonnumerical buttons are usually provided: "#" and "*." They will have more use in the future as computer functions, such as electronic banking, become available in the home. Meanwhile, it should be noted that the tone generators built into the DTMF dial are very precise. The two resultant tones

*Touch-Tone® is a registered trademark of AT&T.

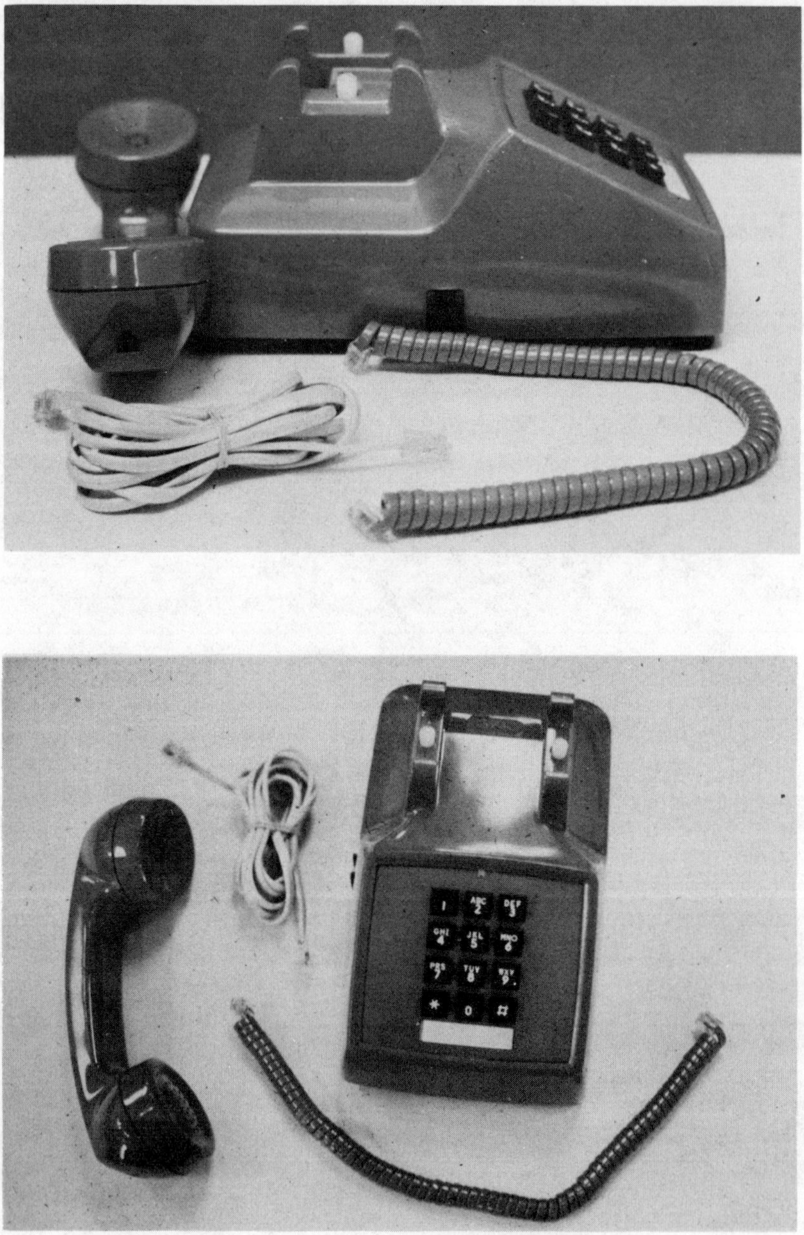

Fig. 4-8. Disassembled views of DTMF telephone.

are precisely related to each other ("twisted") and duplication should not be attempted by the casual experimenter. Federal Communications Commission rules preclude the use of nonapproved items on the telephone line (see "Equipment Registration"). Two views of a DTMF telephone with plug-in cords removed are shown in Fig. 4-8. This is as much disassembly as you should normally attempt.

In summation, DTMF dialing may or may not be available in your area. Check with your local telephone company. DTMF dialing will not work unless your serving central office is equipped to receive it. There is a uniquely different, more melodious dial tone heard from the central office when the central office is equipped to receive DTMF dialing. Some people can tell the difference between rotary and DTMF dial tones by listening to them. You may be able to borrow a friend's DTMF telephone set and see if it works in your home. If you do not have DTMF dialing available, but you want it, the telephone company may be able to provide it by giving you a new telephone number that will accept DTMF dialing. But what if DTMF dialing just is not available from your central office?

There is a way to have DTMF dialing even though DTMF dialing is not offered by your local telephone company. The device that makes this possible is called a *DTMF-to-dial-pulse converter* (Fig. 4-9), and quite often is sold as a replacement for your present

Fig. 4-9. DTMF-to-dial-pulse converter, with replacement DTMF dial.

DTMF dial. It allows you to push-button dial, then converts the signals to standard dial pulses and sends these pulses over your telephone line. The pulses appear to your central office just like standard rotary telephone dial pulses. The converter is usually available as either a conversion for your DTMF set or as a complete new telephone. With it you can push the buttons very fast, but the device will not transmit the dial pulses to the central office any faster than a standard rotary dial. So you are really not able to dial faster, but it may be more convenient for you. This device is particularly helpful for some handicapped people.

DTMF DIALING—TECHNICAL EXPLANATION

DTMF tones are generated by a double-frequency transistorized oscillator. Oscillator power is supplied by the central office over the tip and ring conductors.

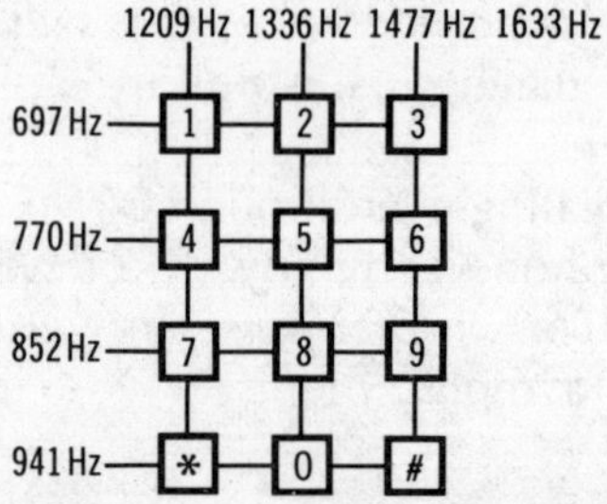

Fig. 4-10. Relationship between push-button numbers and DTMF frequencies.

Two frequencies are generated by depressing any one push button. One frequency will be from the "high" group: 1209,1336, 1477, and 1633 Hz (the 1633-Hz frequency is only seen on 16-button dials used in specialized applications). The second frequency will be from the "low" group: 697, 770, 852, and 941 Hz. The drawing in Fig. 4-10 illustrates the relationship between the push-button numbers and the DTMF frequencies.

In addition to the generation of the two frequencies, a switch is operated that reduces the volume of the transmitted frequencies heard in the receiver, applies bias voltage to the dual-frequency oscillator, and opens the path to the transmitter.

5

The Handset

The handset (Fig. 5-1) provides a convenient mounting for the transmitter and the receiver of your telephone set, and additionally performs the function of holding the cradle switch depressed when your telephone is not in use.

Both the transmitter and receiver may be removed by unscrewing their respective caps. The caps loosen counterclockwise, but

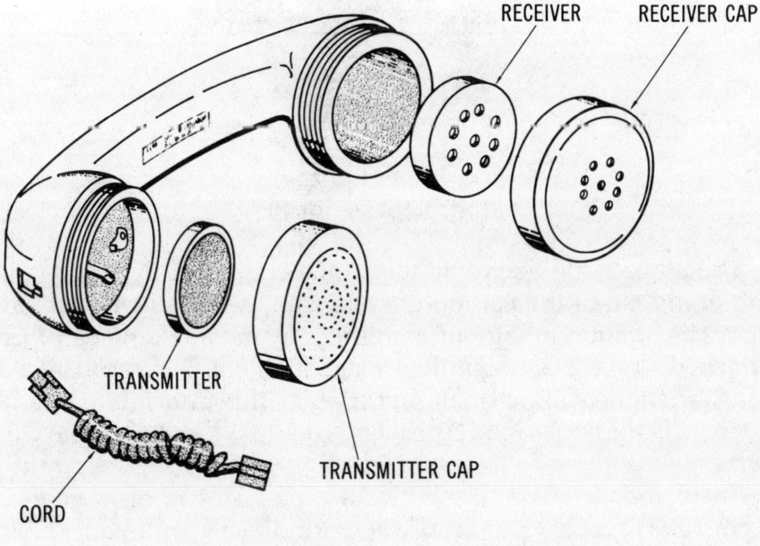

Fig. 5-1. Exploded view of telephone handset.

may be very tight. Telephone company installers may use a special tool to tighten the caps. This is particularly true of handsets used with coin telephones. The tool grips the cap much like an oil-filter wrench grips the oil filter on your automobile.

Unless you are changing handset cords on a nonmodular-cord-equipped handset, you probably won't need to get inside the handset.

THE TRANSMITTER

For many years the transmitter (Fig. 5-2) has been a variable-resistance device, with the variations in resistance being caused by variations in the amplitude of the speaker's voice. As the amplitude increases, the diaphragm causes a compacting of small carbon granules inside the transmitter. The resistance measured through the carbon granules (which is the same as saying the resistance of the transmitter) decreases as the carbon granules are compacted.

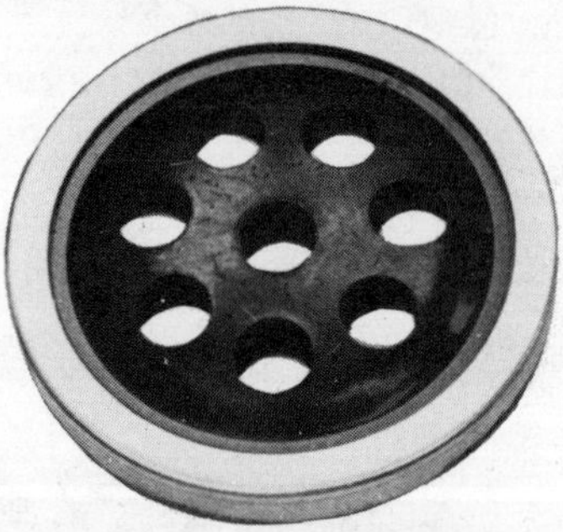

Fig. 5-2. Telephone transmitter.

The newest, all-electronic telephone sets use a dynamic transmitter instead of the carbon-granule type. A dynamic device generates a variation in current amplitude by moving a piece of iron attached to the diaphragm in a magnetic field. The resultant current variations will be small, but usable. This principle is used in audio microphones and speakers such as those in your hi-fi system.

The carbon-granule transmitter will pass a band of frequencies from 400 Hz to 3400 Hz, and will work directly with your serving central office. The dynamic transmitter will pass a much wider range of frequencies—up to 15,000 Hz. But this transmitter requires additional amplification in order to transmit to the serving central office. Hence, dynamic transmitters are only used with electronic networks that include amplifiers, or with separate amplifiers installed in the handset or other suitable location.

THE RECEIVER

The receiver (Fig. 5-3) is a dynamic-type device. Variations in the input voltage cause a changing magnetic field to be generated around a coil of wire. The diaphragm moves in response to the changing magnetic field, pushes against the surrounding air, and your ear "hears" the resultant sound.

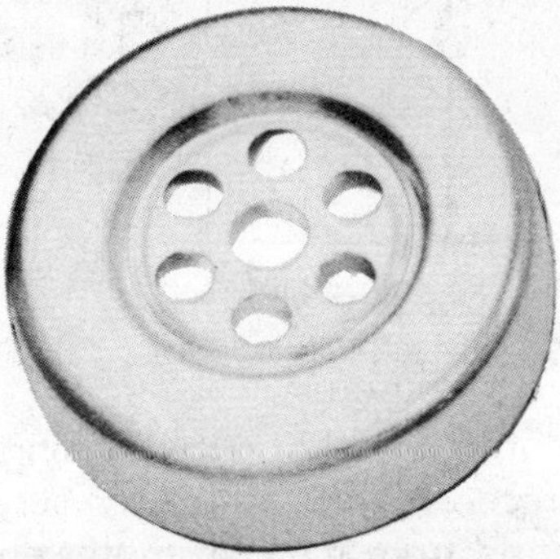

In electronic-type telephones it may be possible to use the same dynamic device for both transmitter and receiver. This helps manufacturers and suppliers by reducing the number of unique major parts in a telephone set, assuming the electronic network contains a very inexpensive amplifier as part of its circuit design. This also permits a wide-frequency-response telephone set to be easily manufactured. The "hi-fi" telephone is thus possible!

To prevent very loud sounds from damaging your ear and possibly the receiver, a volume limiter (Fig. 5-4) is used. The volume limiter consists of a varistor shunted across the receiver terminals. Varistors have a low resistance to loud sounds and a high resistance to soft sounds. Hence, loud sounds will be shunted away from the receiver and soft sounds will be allowed to get to the receiver.

(A) Photograph.

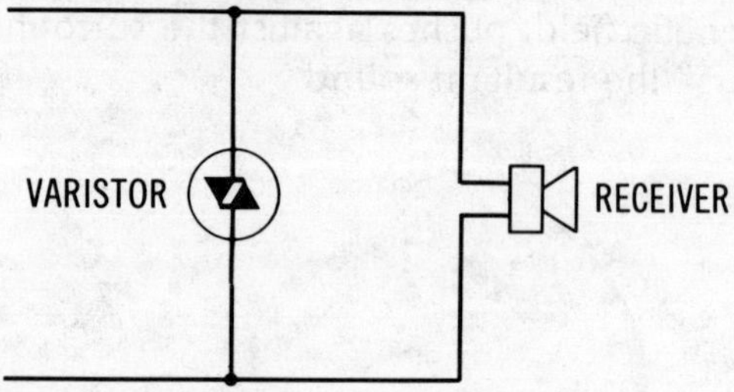

(B) Schematic.

Fig. 5-4. Volume limiter.

Sometimes there is also a wad of cotton found inside the handset, usually near the receiver end. The cotton serves as an acoustic barrier to prevent sound from the transmitter end of the handset traveling inside the handset to the receiver end, and then to your ear. Some handsets do not use a cotton wad, but incorporate some other means of isolation. If you find a wad of cotton in your handset, do not throw it away.

6

The Network

The network inside your telephone includes a circuit that divides the two conductor paths between your telephone and your serving central office into two paths having two conductors each (see Fig. 6-1). One pair of the resultant conductors goes to the transmitter (microphone) and the other pair goes to the receiver (speaker). A small amount of your voice signal is allowed to go from the transmitter to the receiver to let you know how loud, etc., you are speaking. This is called *sidetone*. Screw or spade-type terminals are provided for interconnection points for the components inside the telephone. The ringer capacitor may also be included as a physical part of the network.

The two shunt varistors, V1 and V2 (see Fig. 6-2), produce increased losses on short loops where the voltage from the central office forces excessive current through the telephone. The effective resistance of the varistors decreases as the voltage across them increases; hence, on long loops where the cable resistance limits the current (and thus, the voltage), the varistors present a higher resistance.

Transmitter current variations are in opposite phase in hybrid coil windings TA and TB. Induced voltages in winding TC are also in opposite phase and the resultant voltage is opposed by the voltage produced across the 68-ohm resistor. Hence, the sidetone (signals produced in the receiver by the transmitter) is low and little power is lost in the receiver.

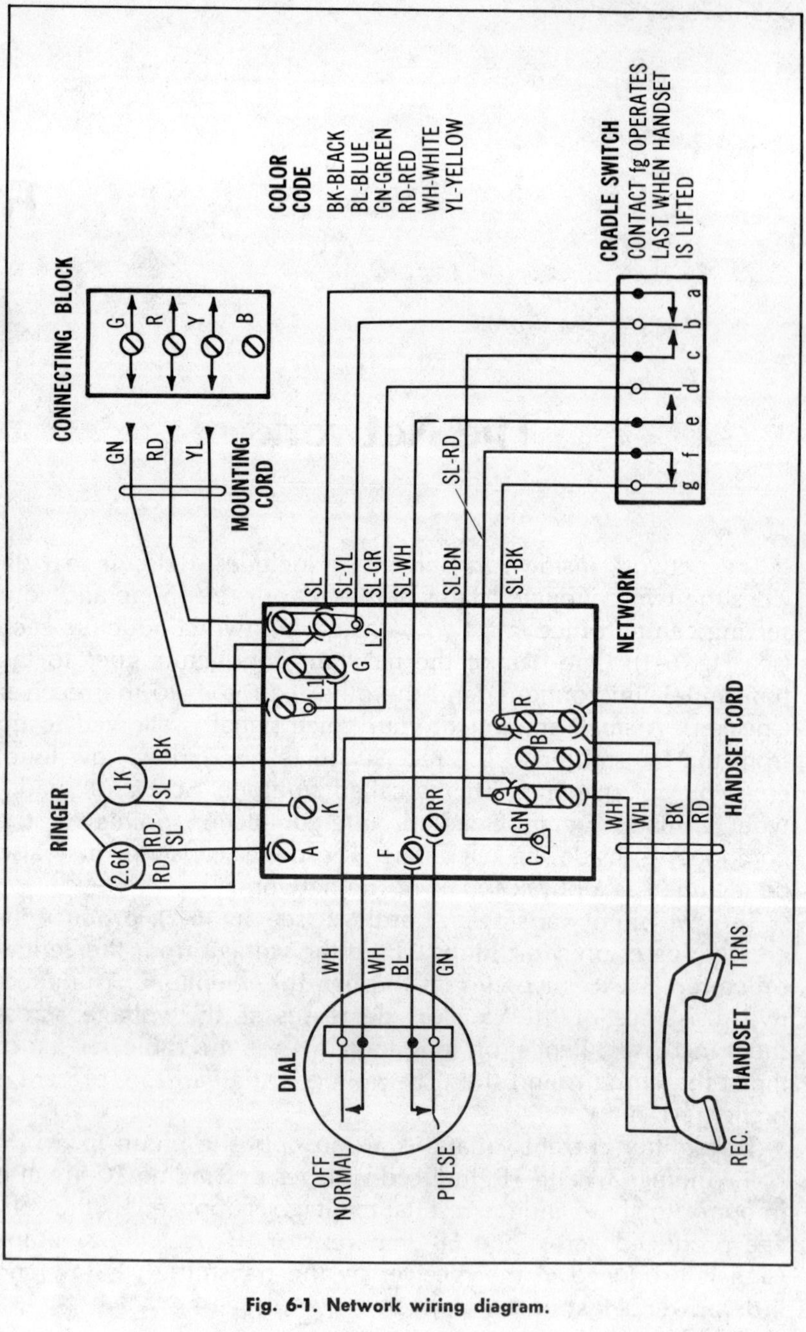

Fig. 6-1. Network wiring diagram.

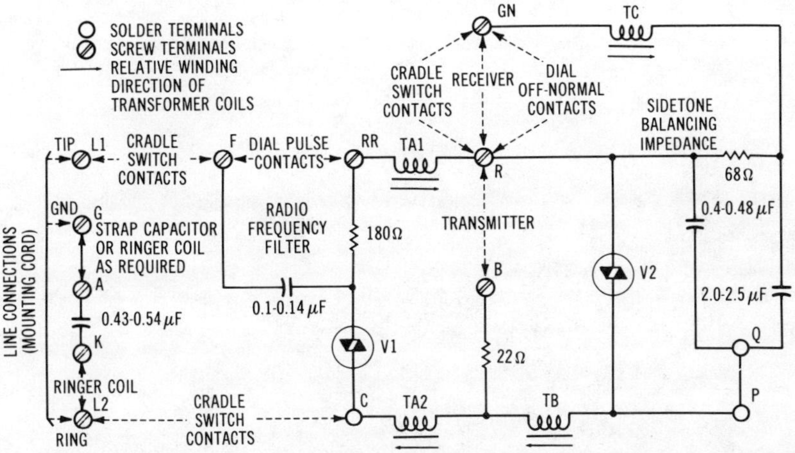

Fig. 6-2. Network circuit diagram.

The radio-frequency filter suppresses high frequencies that result from dial pulsing and could cause local radio interference.

Note the screw terminal designations. From them you can tell where to attach wires when installing or changing components in your telephone.

7

Ringing

SINGLE-PARTY RINGING

Single-party ringing (also called *straight line, Type A,* and by other designations) is supplied by the central office to your telephone set. This ac ringing signal is usually 20 Hz at a voltage ranging up to 150 volts. This voltage will shock you if you have your body, fingers, etc., across the telephone wires during ringing! Normally, the ringing voltage is applied between two of the wires going into your telephone set—the "tip" and the "ring." Since the green wire (the "tip") is also grounded, you can get shocked by being grounded and touching the red wire (the "ring") during ringing. This voltage is present when your telephone is ringing and is not present when your telephone is not ringing.

MULTIPARTY RINGING

Multiparty ringing can use any of a variety of frequencies between 15.3 and 68 Hz, with a peak voltage up to 150 volts. If you live in an area that is phasing out multiparty service, sometimes (rarely) the telephone company will give you private-line service but supply you with a ringing frequency normally used for multiparty ringing. The electromechanical ringer (the bell) in your telephone set is designed to accept and respond (ring) at only one or two frequencies. For example, it may ring for $16\frac{2}{3}$- and 20-Hz ringing, but not any higher-frequency ringing. If you purchase a telephone with a ringer designed for 20-Hz ringing, and the telephone company supplies 50-Hz ringing, your telephone will not ring.

To resolve this potential dilemma, you should first call your local telephone company and ask what type of ringing it is supplying to you. When you have this information, you can then purchase a telephone that is equipped with the proper ringer.

FCC LABELING REQUIREMENT

How can you tell if the telephone set you are purchasing has a ringer that will work in your home? In the words of the FCC:

> "Registered terminal equipment and registered protective circuitry shall have prominently displayed on an outside surface the following information in the following format:
>
> Complies With Part 68, FCC Rules
> FCC Registration Number ________
> Ringer Equivalence ________
>
> Registered terminal equipment and registered protective circuitry shall also have the following identifying information permanently affixed thereto:
>
> (1) Grantee's name;
> (2) Model number, as specified in the registration application; and
> (3) Serial number or date of manufacture."

The "ringer equivalence" contains two pieces of information: the ringer equivalence number (REN) and a letter. The REN will be explained in the next section. The letter following the REN is what we are concerned with. It tells us the frequency that the ringer inside the housing will respond to. Table 7-1 associates the letter with its frequency.

All of this information is usually not needed. What it all boils down to is this: The telephone company supplies Type A ringing for almost all single-party service. Just look for the letter "A" after the number and the telephone will most likely work correctly on your single-party line. This is a lot of information just to tell you

Table 7-1. Range of Compatible Ringing Frequencies

Ringing Type	Ringing Frequencies (Hz)
A	20 ±3 and 30 ±3*
B	15.3 to 68.0
C	15.3 to 17.4
D	19.3 to 20.7†
E	24.3 to 25.7
F	29.3 to 30.7†
G	32.6 to 34.0
H	39.2 to 40.9
J	41.0 to 43.0
K	49.0 to 51.0
L	52.9 to 55.1
M	53.8 to 61.2
N	65.4 to 68.0
P	15.3 to 34.0

* Normal single-party ringing.

† Frequency-selective ringing. Requirements at these frequencies, which are identical to Type A frequencies, are not consistent with the Type A requirements. Equipment intended for use both on Type A facilities and facilities using frequency-selective ringing must comply with the requirements on Types A, D, and F independently.

to look for the letter "A," but now you know why the letter "A" is there.

THE RINGER EQUIVALENCE NUMBER (REN)

The telephone company only supplies a finite amount of ringing power. It only guarantees to supply you with enough power to ring five standard electromechanical ringers (see Fig. 7-1). This is called five ringer equivalence numbers (REN). Each ringer requires one REN to make it ring. You can compare this to an automobile engine. If the engine has only sufficient horsepower to carry five people up a hill, adding a sixth person will cause the car to stall. Adding a sixth ringer to your telephone line can result in none of your telephones ringing.

There is an easy solution. All telephones that you can purchase are required to specify their REN requirement. If the REN number is 2 on your telephone, you can only wire two of them to your line, plus one standard 1-REN ringer (see Fig. 7-1):

2 REN/telephone × 2 telephones = 4 REN
1 REN/telephone × 1 telephone = 1 REN

total = 5 REN, the maximum allowed

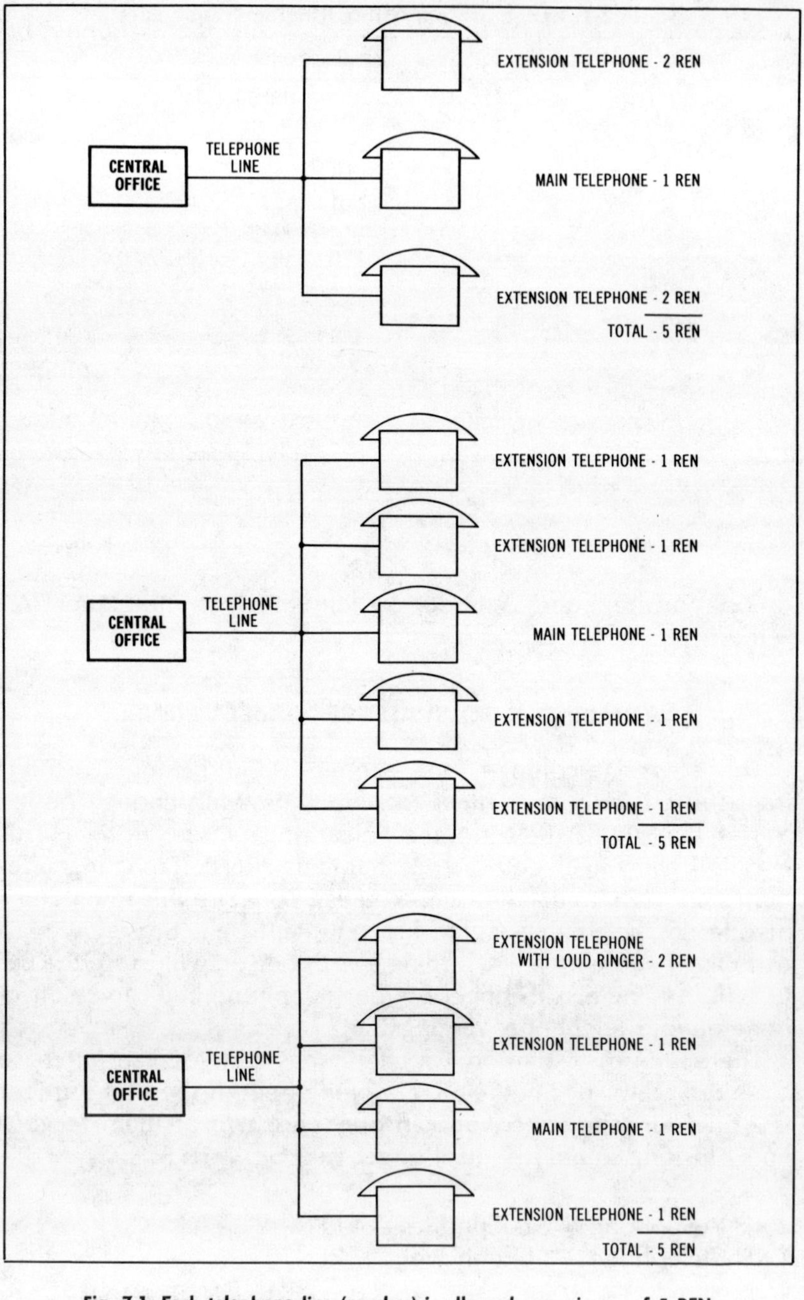

Fig. 7-1. Each telephone line (number) is allowed a maximum of 5 REN.

Conversely, some electronic ringers in some telephones have a REN number as low as 1/10 REN. Logic tells us that we should be able to successfully ring 50 of these ringers:

$$1/10 \text{ REN/telephone} \times 50 \text{ telephones} = 5 \text{ REN, the maximum allowed}$$

You don't have to use up all 5 REN of power, just as your car doesn't need all of its power if it is only carrying one person up the previously mentioned hill. The extra power is not wasted, it just isn't used.

RINGERS

Special ringers are available that are extra loud, sound chimes instead of bells, play tunes, etc. If you are interested in something different from the regular telephone "bells," call your telephone company to see what it has to offer. Then shop around at independent suppliers and see what they have to offer. Compare and make your decision. Your "ringing" telephone might become enjoyable to hear!

8

Talking

When you speak into your telephone, it converts what you say into electrical signals and transmits them along the wires to the distant telephone set. The distant telephone set converts these electrical signals back into speech. When the other person talks to you, the process is reversed.

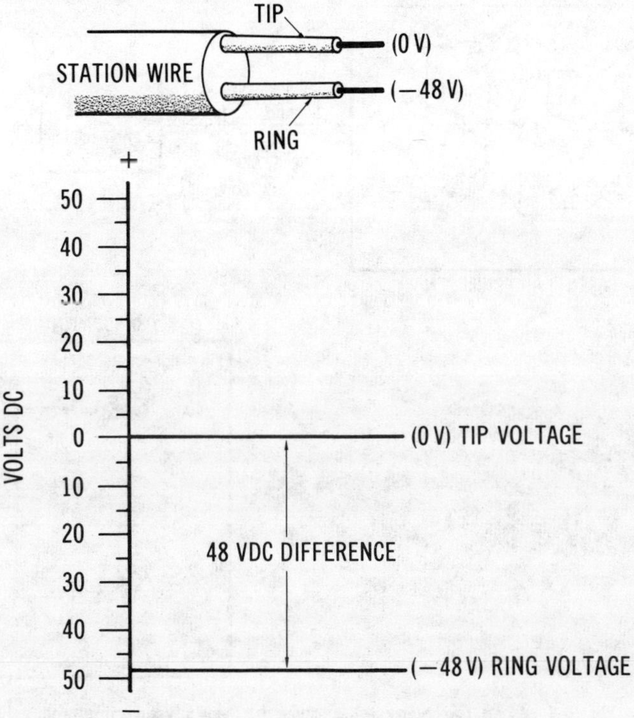

Fig. 8-1. Voltage supplied by telephone company across tip and ring wires.

In order to accomplish this, the local telephone company supplies a nominal 48 volts dc across the tip and ring wires. The tip (the green wire) is at ground potential, or zero volts. The ring (the red wire) is at negative 48 volts. Therefore, the tip is more positive than the ring (see Fig. 8-1). If this seems confusing, just keep in mind that there are 48 volts dc across the tip and ring wires. There is only a limited amount of power available in the 48 volts; approximately 21 to 35 milliamperes of current, which would provide a maximum power of about 1.0 to 1.5 watts.

The Bell System proved some time ago that between 21 to 35 milliamperes of current was required to properly operate a telephone set. Your telephone will work satisfactorily at values close to these. Note that if you are talking to someone on your telephone, and someone else at the distant end picks up an extension telephone, the volume of both persons at the other end is reduced. Grandparents know that when their son, his wife, and all their

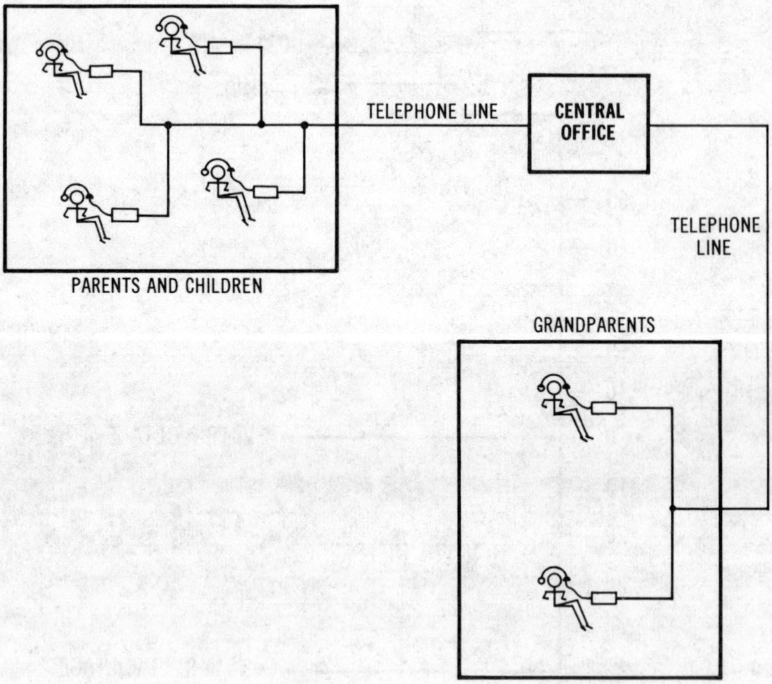

Fig. 8-2. When too many telephones are being used simultaneously, the volume will be very low.

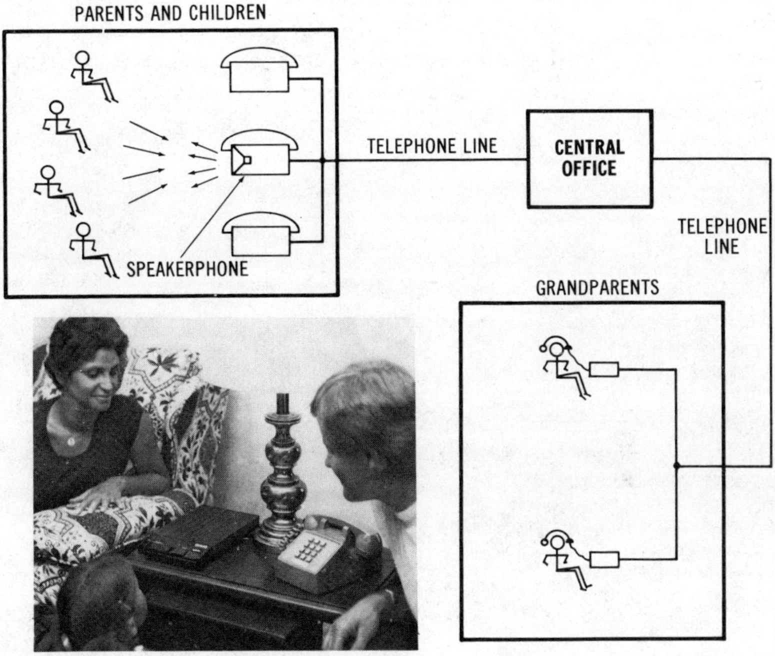

Fig. 8-3. Family using speakerphone.

children use different telephones, Grandma and Grandpa can barely hear any of them (Fig. 8-2).

There is a number at which there will be insufficient power available to operate two or more colocated telephone sets simultaneously. This number is a function of the distance your home is from the central office, the type of telephones you want to use simultaneously, and other variables that would be almost impossible for you to determine. If you only use one or two telephones at a time you will have no problems. Some new types of telephone sets overcome this power limitation by using much less power per set than the older types. However, these sets are not generally available to the public.

The result of this is that while you may have telephones installed in every room, you can probably use only two or three of them simultaneously. This is not normally a problem. If more than two of you must use your telephones you should consider purchasing a "speakerphone." With this system, everyone in the room can talk (Fig. 8-3).

9

Home Wiring

If the telephone company will allow you to wire your own home (or business, etc.), read on. You should check with the telephone company first, and also check for any applicable local wiring codes. After you have done this you are ready to proceed.

The telephone company primarily uses No. 22 AWG size wire, with two, three, or four conductors; with four being preferred. Suitable wire can be purchased at most radio supply stores (Fig. 9-1).

A minimum of two conductors are required to make your telephone work. These are the "tip" and the "ring" conductors. Normally, the ring conductor will be red and the tip conductor will be green. The third conductor will be yellow, if there is one, and may be used as a ground wire, or in conjunction with the fourth conductor to provide low-voltage current to light a dial lamp, etc. The fourth conductor will be black, if there is one. If you find more than four conductors in your existing home wiring, you have to determine which two are used for the tip and ring and use those conductors throughout your home. Once a particular color conductor is established as a tip or ring, that color will be associated with that particular conductor throughout your home.

Assuming your home already has at least one telephone installed by the telephone company, you can easily determine which conductor to use for tip, ring, etc., by looking at the receptacle. Simply remove the cover by removing the screw on the front of the cover. Remove the receptacle from its mounting box by removing the one or two mounting screws holding the receptacle to the frame. Now you can see which color conductors are wired to which positions on the back. Wire your new receptacles exactly like the one you have. If you have a 42A block, remove the cover by removing the screw in the cover and observe which color conductors are being used. (See "Receptacles and Connecting Points (Blocks)" for an identifying picture.) The green wire will be the tip and the red wire will be the ring. On your new receptacle there will either be a raised letter "t" for *t*ip, or else a raised letter "g" for green, and so forth for the *r*ing (*r*ed), the *y*ellow, and the *b*lack.

Fig. 9-1. Station wire.

If your home has no existing wiring and you are going to do all the wiring yourself, there are two possible ways to proceed. The first method is the preferred one. Your local telephone company will probably provide this type of installation for you if it knows you are going to wire your home yourself. This method, or a close variation of it, will probably be the only one allowed in the future if the telephone companies and the FCC all reach a common agreement as proposed.

A major concern of the telephone industry is that you will inadvertently disable the station protector. Lightning could then damage your home and the telephone company facilities. To prevent this, the telephone companies have proposed a method of interconnection to keep you out of the station protector (Fig. 9-2).

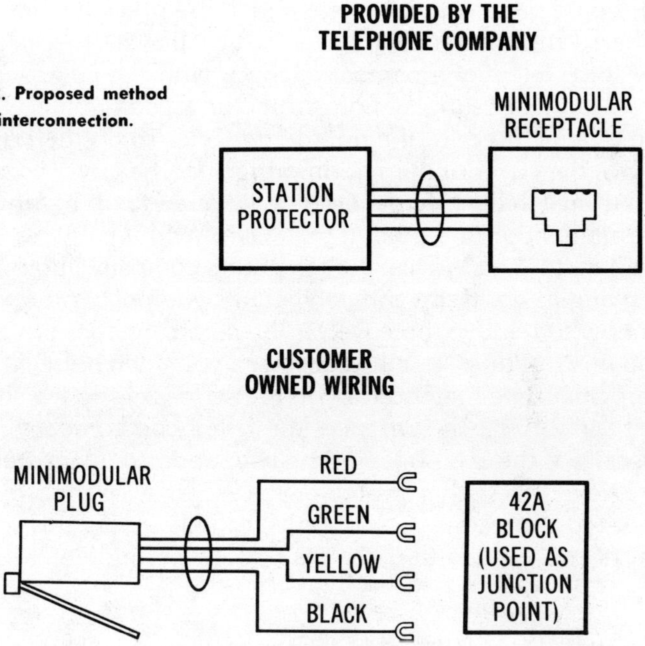

Fig. 9-2. Proposed method of interconnection.

Note that the telephone company proposes to install a standard minimodular receptacle inside your home, close to the station protector. You will then use a minimodular plug-ended cord to connect to the telephone company facilities. You could install a 42A block, or equivalent, near the telephone company receptacle, and use a short "plug/spade" cord to connect the two. Then wire from the 42A block to the rest of the receptacles in your home. With this method, the 42A block is used merely as a junction point for your home wiring and may not have a telephone set next to it. Not only does this method keep you away from the station protector, it also provides one point—the telephone company receptacle—where your home wiring can be easily disconnected from the telephone company facilities when checking for trouble on the line.

After you insert the minimodular plug into the telephone company receptacle, the red wire at the spade end of the plug/spade cord will be the ring, the green wire will be the tip; the yellow and black wires will be likewise identified if they are to be used. All wires from your 42A block can be coordinated throughout your home once you have determined the preceding information.

The second method, and one which may not be allowed by your local telephone company (check with the company to see), is to use the existing station protector to determine what color wire is being used for the tip, the ring, etc. To do this you have to remove the cover of the station protector. The cover usually snaps off, but may have a screw holding it. Screws will be obvious and accessible if they are used. If you have any difficulty removing the cover, quit trying. Advise the telephone company about what you are trying to do. If the company grants you permission, it will tell you how to remove the cover. If the company does not allow you to do this, but does permit you to wire your own home, it will have to provide some method for you to use to identify which conductor is the tip, the ring, etc. Let the telephone company help you, especially if there is no fee. If the telephone company permits you to remove the station protector cover, do so and note the three large studs (threaded ends of bolts) called *binding posts* (Fig. 9-3). One of these studs (usually the one in the middle) will be con-

Fig. 9-3. Station protector.

nected to a large copper wire that can be traced to an electrical ground. The ground wire may connect to a water pipe or to a large steel/copper stake that has been driven four or more feet into the ground. This wire is the path that lightning will travel if it strikes your telephone line! Do *not* disconnect this large wire! If you need a ground wire to your telephone set, connect the yellow conductor of your home wiring to this stud. Now the yellow wire (or whatever color happened to be used if you have more than four-conductor home wiring) is at ground potential. If your telephone set does not require a ground, do not bother connecting the yellow wire. The instructions provided with your new telephone set will tell you if you require a ground.

This is a good time to explain how to properly connect a conductor (bare wire) to a stud-type mounting (binding post) (Fig. 9-4). Note the use of small flat washers between conductors. This helps ensure a good, secure connection for all conductors. In the station protector photograph (Fig. 9-3), you can see the flat washers between the conductors and between the nut and the top conductor. Now let us return to our discussion of the station protector.

The other two studs in the station protector will have a green (tip) and a red (ring) wire connected to them from the telephone company's outside line. If no red and green wires appear, the

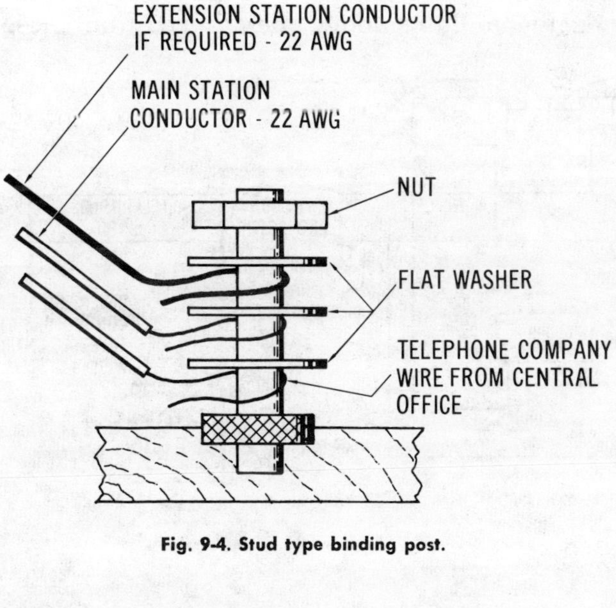

Fig. 9-4. Stud type binding post.

Table 9-1. Wire Connection Guide

Wire Color	Color of Existing Telephone Wire Terminated on Station Protector
Red	Red or blue with white stripe.
Green	Green or white with blue stripe.
Yellow	Yellow or orange with white stripe.
Black	Black or white with orange stripe.

telephone company has used different color-coded wire and you will have to either measure or guess as to which wire is which. Note that there is no stud to connect the black conductor to. See Table 9-1 for a guide.

All that is really important is to wire all receptacles consistently. If you put the tip conductor on one position in a receptacle in the home, put it on all receptacles in the same manner. If you wind up with the tip and ring conductors reversed, all you have to do is reverse them at the station protector and the entire home wiring system will be corrected (Fig. 9-5).

If you are merely adding wiring to a previously wired home, take an existing receptacle apart, determine which conductor goes where, and wire your new receptacles identically. Some new telephone sets are being manufactured with a device called a *polarity guard* that makes the telephone insensitive to the polarity of the tip and ring conductors. These telephones allow the tip and ring conductors to be randomly interchanged from receptacle to

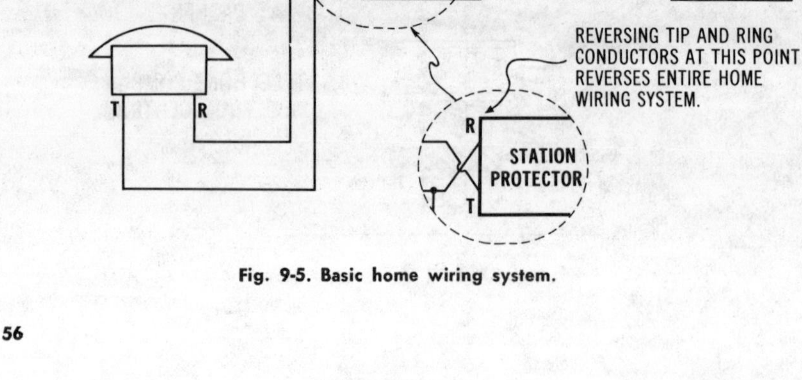

Fig. 9-5. Basic home wiring system.

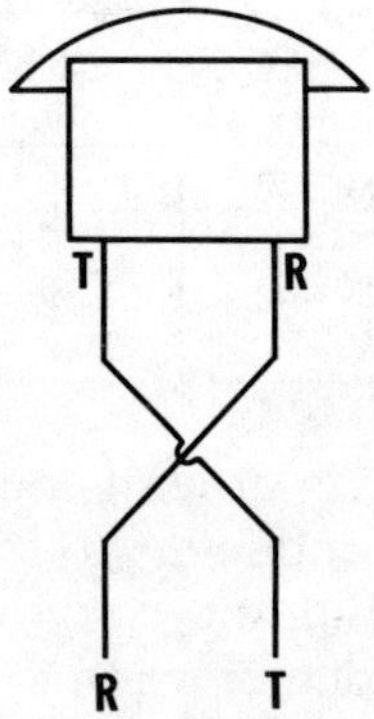

Fig. 9-6. Reversing tip and ring conductors at telephone.

receptacle with no problems resulting. As a good installation practice, coil up about a 10-inch length of each unused conductor inside the protector and receptacle boxes away from the wires you used.

What happens if you accidentally get the tip and ring conductors reversed? No damage will result, but you will not be able to dial out with a DTMF telephone. If you are sure you can use DTMF dialing on your line, but when you depress a button the dial tone does not stop, reverse the tip and ring conductors (Fig. 9-6).

A rotary dial telephone will work with the tip and ring conductors reversed, but when you dial, your telephone ringer may "ding" in time to the dial pulses. This is called *bell tap* (not *Bell System* tap, but *ringer-bell* tap). Reversing the tip and ring conductors will cure this condition.

A note here is appropriate. You can always change a telephone to make it work with an incorrectly wired receptacle by reversing the wires inside the telephone set. But that is considered poor practice and should *not* be done. Also, the telephone will then work only with a similarly miswired receptacle. If you wire everything correctly the first time, you will save yourself future problems on any additional wiring you may do.

WIRING FROM PROTECTOR TO RECEPTACLE

Use No. 22 AWG wire with at least three conductors. If you foresee any devices being used other than standard telephone

Table 9-2. AWG Size Versus Conductor Diameter

AWG Size	Nominal Diameter		Electrical Resistance (Ohms/1000 ft @ 20°C)
	Inches	Millimeters	
26	0.0159	0.404	83.33
24	0.0201	0.511	51.89
22	0.0253	0.643	32.58
19	0.0359	0.912	16.29

sets, wire with at least four conductors (you can use the yellow and black conductors to supply power for telephones with night lights built in). And if you want to cover 99% of all eventualities, use six-conductor wiring. Although some telephone companies may use No. 24 AWG or No. 26 AWG station wiring, you should still use the No. 22 size to be safe. It never causes problems to use larger diameter conductors than necessary. It *can* cause problems to use smaller diameter conductors, primarily due to the higher electrical resistance of the smaller conductors. Table 9-2 shows that wire diameter is inversely related to its AWG (American Wire Gauge) size; No. 22 AWG wire is larger than No. 24 AWG, etc.

Plan your installation. If you are wiring in a home under construction, you will not have many problems. In existing construction, you will have to plan the route of your wiring more carefully.

Telephone sets are wired in parallel. That is, all tip wires are connected together, all ring wires are connected together, all yellow wires are connected together, etc. You can make individual multiconductor cable runs from each receptacle to the station protector. This is called *home-run* wiring (Fig. 9-7A). Or, you can run one cable from the station protector to the closest receptacle, then from that receptacle to the next closest receptacle, and so forth. This is called *multipled* wiring (Fig. 9-7B). Just keep all the same color wires together, and attach them to the same terminal in each receptacle.

How far away from electrical wiring, pipes, etc., should you stay to be safe? Table 9-3 shows the clearance that is generally acceptable. Your local building codes will be your best authority and should be consulted before beginning your wiring.

It is considered good practice to leave sufficient slack in the wire to relieve strain at terminals, to provide for rearrangements, and to give sufficent clearance for proper operation of nearby equipment. Also, be sure to protect the wire from potential me-

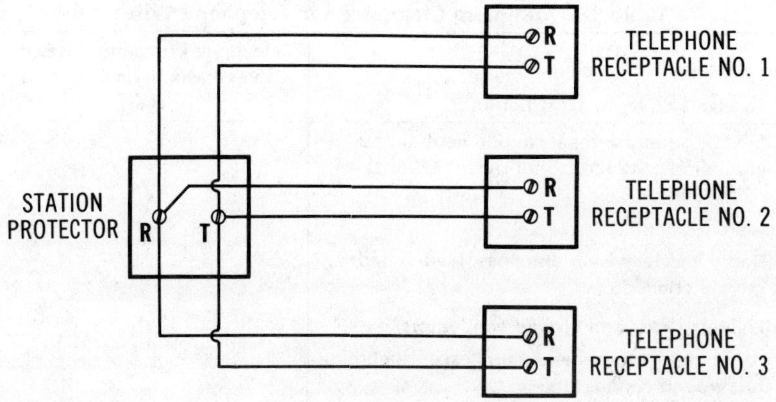

(A) "Home-run" wiring.

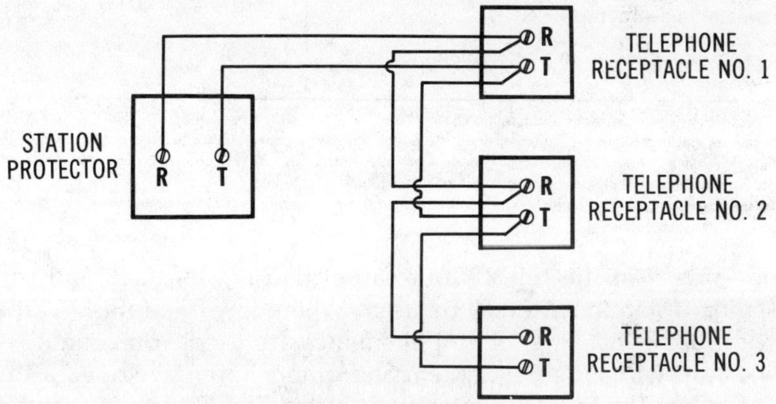

(B) Multipled wiring.

Fig. 9-7. The two basic methods of telephone wiring.

chanical damage by enclosing it in conduit or other protective coverings.

UNDERCARPET CABLE

What if you have a concrete floor and wish to install a telephone on a desk in the center of the room? The ideal telephone station installation wiring should be out of sight and easy to install. With the introduction of "flat" station cable you can now run the necessary number of conductors under rugs with little or

Table 9-3. Minimum Clearance for Telephone Wires

Foreign Facility or Obstruction	Telephone Company Wires or Cables—Minimum Clearance (Inches)*
Electric supply wires including neutral and grounding conductors (not over 750 volts).	
Open	4
In conduit	2
Radio and television antennas, lead-in and ground conductors.	4
Lightning rods and ground conductors.	72†
All foreign ground conductors except lightning rod ground conductors.	2
Neon signs and associated wiring.	6
Metallic objects—pipes (gas, water, oil, sewer, steam) and structures.	2‡
Wires or cables of another communication system.	2

*Clearances apply to crossings and parallel runs.

†If this separation cannot be obtained, bond the telephone grounding conductor or ground electrode to the lightning rod grounding conductor or ground electrode with a No. 6 AWG copper ground wire. With this provision, a minimum clearance of four inches is acceptable.

‡Increase to minimum three-inch clearance from steam or hot-water pipes, heating ducts, and other heat sources.

no "ridge" in the rug. If your carpet pad is very thick and soft, standard station wire may be used. Where low-pile (plush) carpet with rubber backing, or carpet squares are used, you cannot use standard wire. The Bell System and others have developed a thin flat cable. The Bell System cable is called TUCC® and is currently only available with 25 pairs of conductors. Other manufacturers supply smaller pair-count flat cables. When selecting flat cabling, check to be sure that it is designed for your contemplated use. Undercarpet cable has to withstand impact and crushing forces, and any spilled liquids and carpet cleaning chemicals. The total height of the cable should not exceed the generally accepted height of 80 mils (0.08 inch or 2 millimeters). This is approximately as thick as six calling cards. Tape flat cable to the floor with standard PVC (electrical) tape so it will not shift when you reinstall the carpet.

Flat undercarpet cable offers a virtually invisible installation, at low cost, to meet needs that conventional round station cable cannot. Check your local radio supply store for availability. Be sure the flat cable is specified as having flat conductors equivalent to the same size round conductors you are using for the rest of your installation (usually No. 22 AWG). Otherwise, you may find the flat cable conductors are too small to pass the amount of current required by your telephone set; or, conversely, the flat cable conductors are unnecessarily large and awkward to install. If you have to guess, guess on the large side; at least your telephone will work.

10

Receptacles and Connecting Points (Blocks)

You should provide some easily disconnectable connection points between the wiring in your home and the wiring from your telephone set. This should be considered mandatory. It will allow you to easily move your telephone to different locations in your home. Radio supply stores are a good source for these items. Usually, any store that sells telephones also sells blocks, receptacles, wiring, etc.

BLOCKS

The Bell Telephone System uses a "42A block" for nonreceptacle-equipped installations (Fig. 10-1). Other telephone companies use a similar appearing block for this type of installation. By removing the screw in the cover, the cover can be removed. This will expose four screw tie-down points (Fig. 10-2).

Wire the tip conductor from your telephone set to the tip conductor in your home wiring. Do likewise for the ring conductor. Some blocks are coded to indicate which conductor should go where. If you have yellow and black conductors, wire them to their matching conductors. (All we are doing here is matching color to color, unless we have different color conductors. See the "Wire Connection Guide" in Chapter 9.) Replace the cover and you are done. If you are mounting the block, choose some out-of-

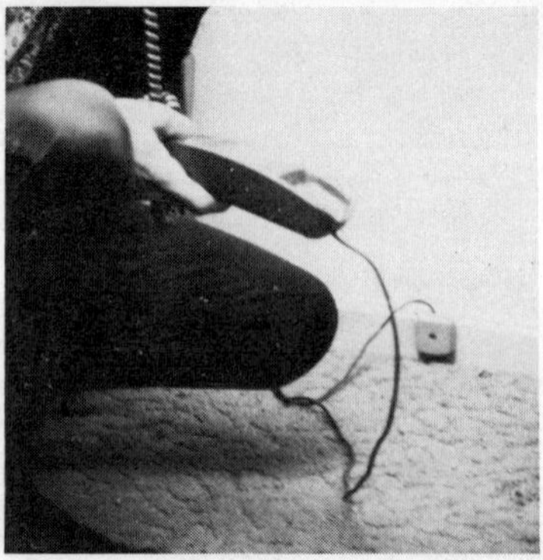

(A) Photograph of installation.

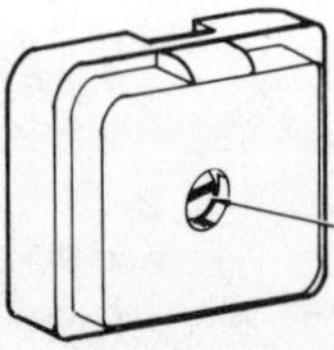

(B) Sketch of cover.

Fig. 10-1. The 42A connecting block.

the-way place that is as convenient as possible and attach it to the surface with screws, then replace the cover. Obviously all blocks can be installed prior to wiring the telephone sets to the blocks.

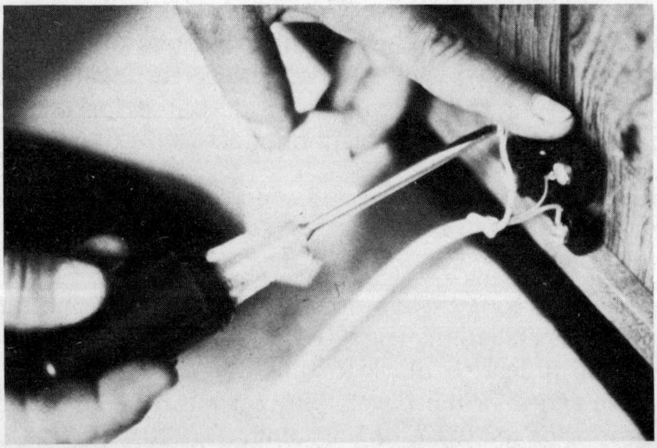

Fig. 10-2. Photograph of 42A connecting block installation with cover removed to show wiring.

FOUR-PRONG RECEPTACLES AND PLUGS

If you want to make your home wiring even more convenient for moving telephones to different locations, four-prong receptacles and plugs can be used. Fig. 10-3 shows top and bottom views of a typical receptacle of this type and two different plugs that will mate with it.

To install this type of connecting point, first dismantle your new receptacle. If there are other receptacles in the home, remove one cover and determine which prongs are wired to the tip, ring, yellow, and black, if used. Wire the new receptacle identically and then mount it. Dismantle the four-prong plug and wire

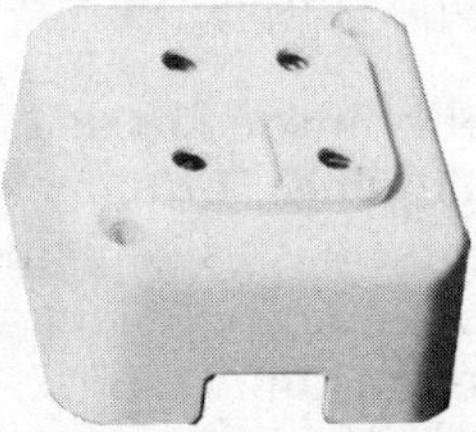

(A) Top view of receptacle.

(B) Bottom view of receptacle.

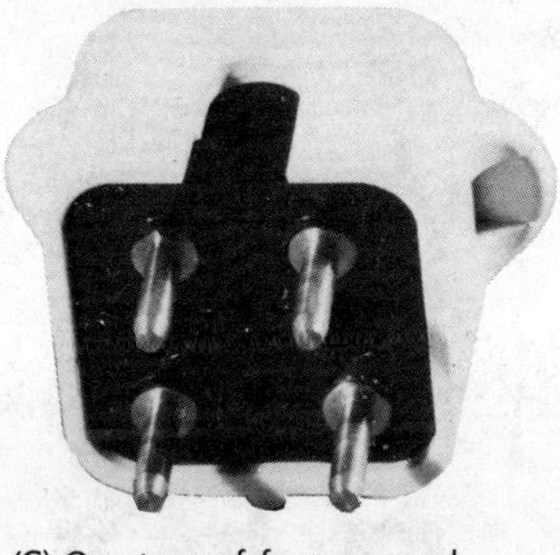

(C) One type of four-prong plug.

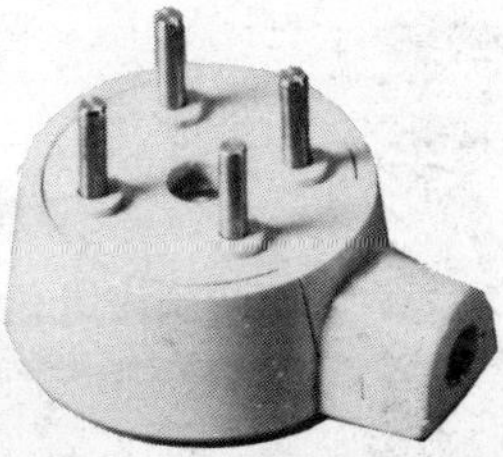

(D) Another type of four prong plug.

Fig. 10-3. Four-prong receptacle and plugs.

it to match tip to tip, etc., with the receptacle. You may find that the receptacle and plug are coded to assist you in attaching the conductors to the correct prongs.

MINIMODULAR RECEPTACLES AND PLUGS

Minimodular receptacles and plugs are the newest method of providing connecting points that are easily disconnected. Fig. 10-4 shows a surface-mount minimodular receptacle being installed, while Fig. 10-5 shows a flush-mount minimodular receptacle and the cord plug that mates with it.

To install this type of receptacle, follow the same procedure as that used for four-prong receptacles. The plug, however, has been permanently attached to the telephone set wiring. There are two types of modular telephone set wiring: plug/spade and plug/plug (Fig. 10-6).

Plug/spade wiring has a minimodular plug on one end and spade lugs on the other. They are preterminated at the factory. You should not try to shorten or splice into the silver-satin colored wiring. This could easily ruin it. Your telephone set may have this type of wiring installed. If not, a meter is needed to determine where each conductor is connected in the plug. Fortunately, the conductors are color coded and the receptacle is usually marked. Inside the telephone set is a group of labeled screw connections. Attach the green conductor to L1 and the red conductor to L2 by following the telephone manufacturer's instructions. Note here that some sets require the yellow conductor (if used) to be at-

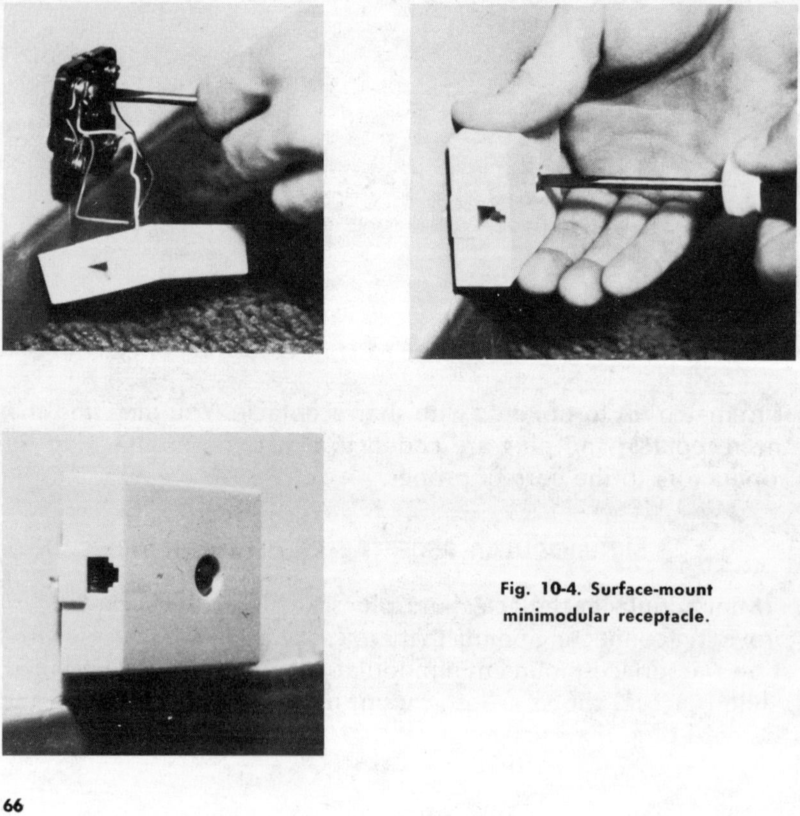

Fig. 10-4. Surface-mount minimodular receptacle.

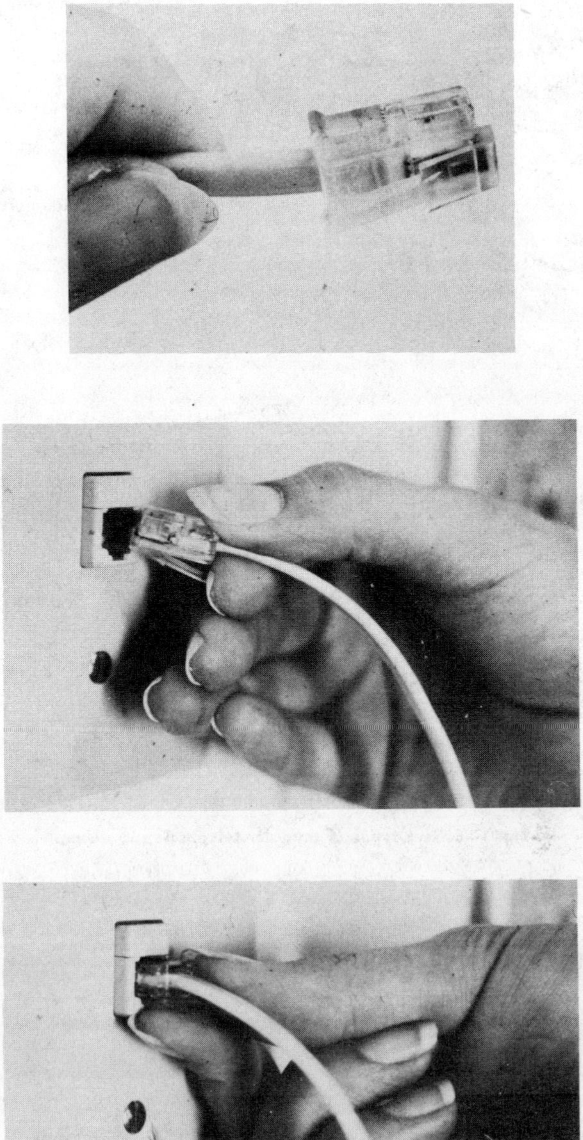

Fig. 10-5. Flush-mount minimodular receptacle and plug.

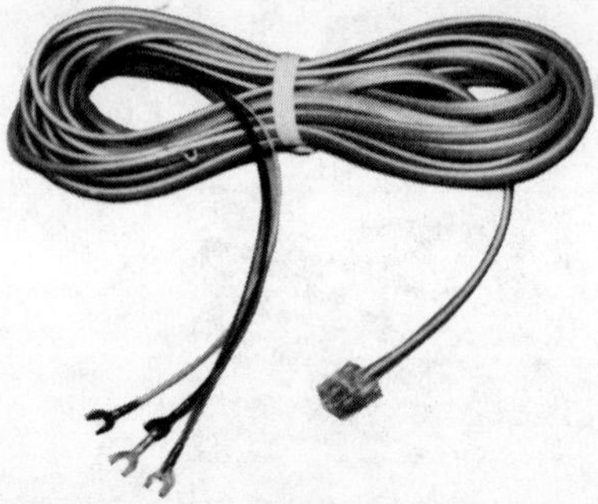

(A) Plug/spade cord.

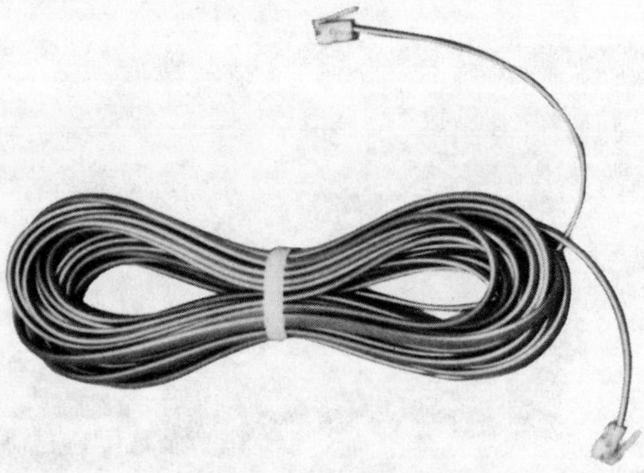

(B) Plug/plug cord.

Fig. 10-6. Two types of modular telephone set wiring.

Fig. 10-7. Conversion jack—42A block to minimodular receptacle.

tached to the same terminal as the green conductor, or ringing may not occur.

Plug/plug wiring requires a receptacle in the side of the telephone set and another receptacle on the wall. Just plug either end of the wiring into the telephone set and the other end into the wall receptacle.

A simple alternative to converting your 42A blocks to minimodular types is provided by using a conversion jack (Fig. 10-7). Installation is accomplished by simply removing the 42A cover by loosening the center cover screw, replacing it with the conversion jack, tightening the cover screw, and plugging in the modular-cord-equipped telephone.

11

Station Equipment

Let us assume you have a completely wired home ready for the telephone sets and devices of your choosing. You now know the difference between (and the availability of) DTMF and rotary dialing for your home. What can you get to further customize your telephone equipment?

The first thing you should do is call your local telephone company business office and determine the types of telephone equipment and services offered by the local telephone company. Note the word *services*. There are many features available in some localities that may be desirable; however, due to the complexity and cost of the required equipment, such features are presently available only from the telephone company. These include:

Audible Signal On Incoming Call—A "beep" tone sounds once in your receiver when you are talking while another person is trying to call you. If you hang up within a few seconds, the second call will be connected to you without the calling party ever knowing that your line was busy.

Camp On—This feature allows you to call, and if the called party is busy, you can hang up; your call will then wait until the called party hangs up, at which time both telephones will ring simultaneously.

Call Forwarding—This feature allows you to transfer all calls coming to your number to another number of your choice by dialing a special sequence of digits. This is very handy if you are expecting an important call but want to go elsewhere (say to the neighbor's for dinner). You cancel call forwarding by dialing another special sequence of digits.

Conference Calling—This feature allows more than two telephone numbers to be connected simultaneously. This is great for telephone meetings.

New features and services appear quite often. Call your local telephone company and inquire as to what features and services are available to you.

12

Specialized Tools

As with any specialized type of work, the telephone business has a number of specialized tools. These tools are used to increase the efficiency of telephone personnel. Although most of the work you are going to do will not require any tools beyond the ordinary ones found around most homes, there are a few items that can prove very helpful and will increase your installation efficiency.

BELL HANGER BIT

This drill bit (Fig. 12-1) is usually 1/4 inch in diameter or larger and has a total length of 12 or 18 inches. You should buy the 18-inch length if possible. With this bit, you can drill all the way through your floor, the floor joists, and the ceiling below. There is a small hole in the center of the bit about one inch from the cutting end. Once the bit has made a hole, do not withdraw the bit. Rather, insert one of the conductors of your wiring cable

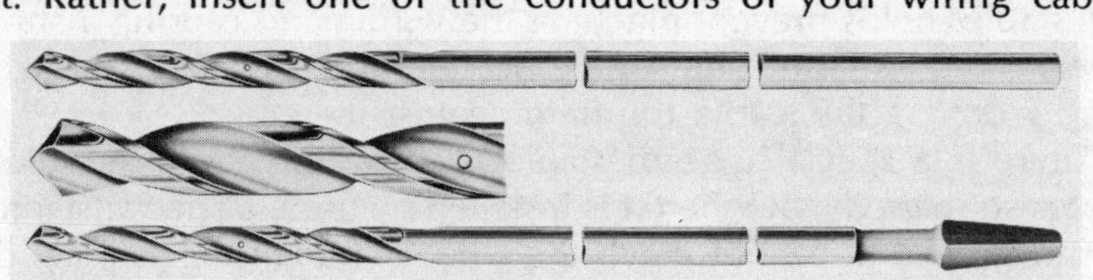

Courtesy Irwin Augar Bit Co.

Fig. 12-1. Bell hanger bit.

Table 12-1. Suggested VOM Measurement Ranges

Range	Purpose
0–100 V dc	Voltage between tip and ring conductors.
0–20 V ac	Dial-light voltage (can be combined with 0–200 V ac).
0–200 V ac	Ringing voltage and perhaps dial-light voltage.
0–50 mA	Current supplied by central office battery.
0–1000 ohms	Continuity tests (any higher resistance range will do).

through the hole and then withdraw the bit. You will automatically pull the entire cable through the hole. You can purchase bell hanger bits from most hardware stores.

VOLT-OHM-MILLIAMPERE METER (VOM)

A volt-ohm-milliampere meter (vom) is an essential tool for making various tests and measurements in any electrical circuit. Since you are only going to use a vom for testing continuity (determining if there is a break in the wiring), or for measuring very basic voltages and currents, a simple inexpensive instrument such as the one shown in Fig. 12-2 is all that you will need. Suggested measurement ranges for your vom are given in Table 12-1.

Follow the manufacturer's instructions when using your vom. Be careful to not drop it! Meters are delicate instruments and must be handled with care.

If you already have a meter, check the condition of the internal batteries before using it. Weak batteries can give wrong readings that will cost you endless time trying to find a trouble that doesn't exist.

INSULATED STAPLES

Insulated staples (Fig. 12-3) provide a convenient method of securing your home wiring to walls, floors, etc. The insulation on the staples helps prevent pinching the wiring and causing a short between conductors. You should be careful to not mash the wiring by driving the staple too hard against it.

There is a special type of staple used by telephone installers that is so shaped that when it is installed by using a special stapler gun (Fig. 12-4) it cannot pinch the wire. Although this is a handy tool, it would not be very practical if you are only going to do a small amount of wiring that involves stapling.

Courtesy Triplett Corp.

Fig. 12-2. A volt-ohm-milliampere meter.

MODULAR CORD AND WIRING TEST SET

I will now describe a simple test set that you can use to test and troubleshoot the wiring in your home. Do not be inhibited by the schematic diagram (Fig. 12-5). The set is easy to build and tests itself.

Note that the test set can be used to test wiring in locations using key telephone systems (kts) and private branch exchanges (pbx). These will probably not be encountered and any reference to them should be ignored. However, for those persons using them I have included the necessary information. All parts can be purchased at most radio supply stores.

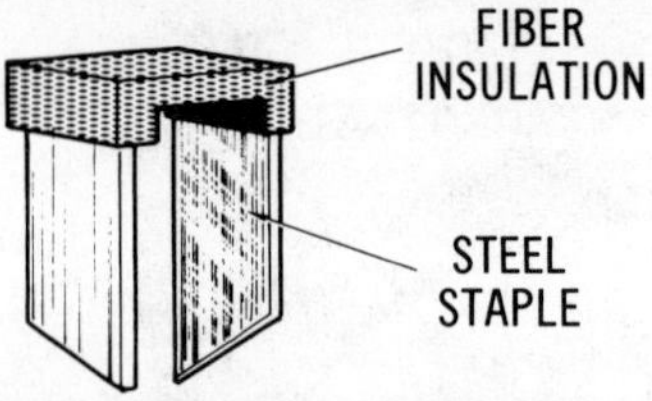

Fig. 12-3. An insulated staple.

Purpose

1. To test modular cords for shorts, crosses, and continuity.
2. To test station wiring for shorts and crosses prior to connection to central office, pbx, or key system battery "dry" circuit.
3. To test station wiring for grounds prior to connection to central office, pbx, or key system battery "dry" circuit.
4. To test station wiring for tip and ring reversals, continuity,

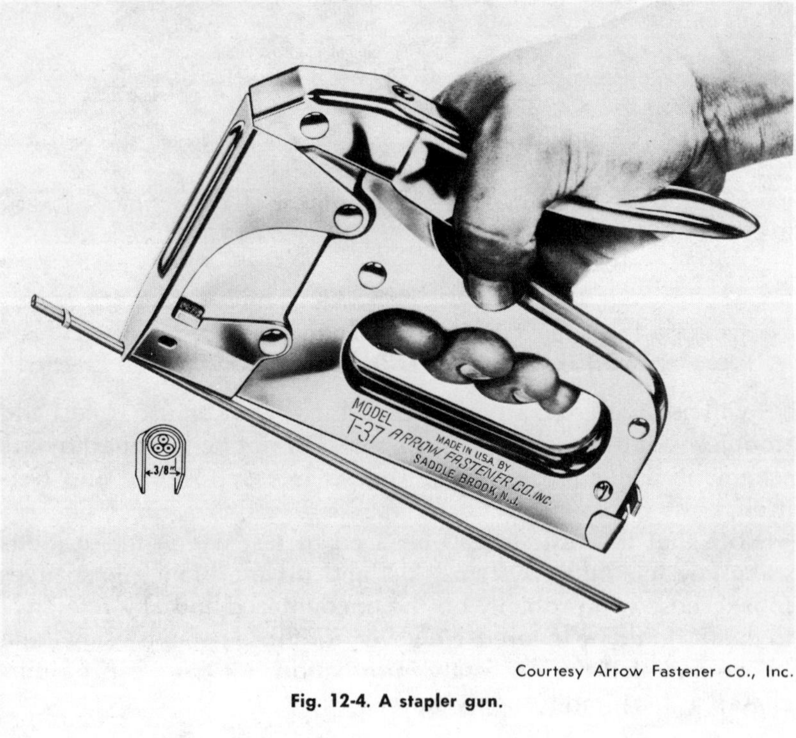

Courtesy Arrow Fastener Co., Inc.

Fig. 12-4. A stapler gun.

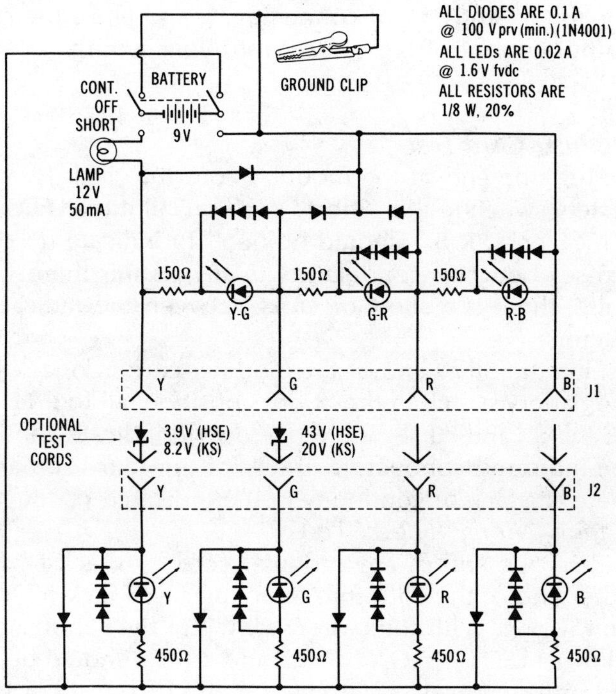

Fig. 12-5. Schematic diagram of modular cord and wiring test set.

minimum voltage available on the tip and ring conductors, minimum voltage (ac or dc) available for lamps on the yellow and black conductors (while connected to central office, pbx, or key system battery "wet" circuit), and lamp power supply (6 volts, 10 volts, or greater, ac or dc). In lieu of the presence of central office, pbx, or key system battery, two 9-volt batteries can be connected across the wiring at a common point, such as the station protector, and a regular modular cord used with the test set to verify wiring continuity and tip and ring reversals. One battery should be connected with the negative terminal to the green wire and the positive terminal to the red wire, and the other battery connected with the negative terminal to the yellow or black wire and the positive terminal to the unused black or yellow wire.

5. To test the condition of the internal battery of the modular cord and wiring test set.
6. The lamp limits battery drain in case of a short between ter-

minals "B" and "Y" of connector J1, and limits the potential output current of the test set into your wiring.

Operation

1. *Modular Cord Test*

   Plug one end of the modular cord into jack J1. With the battery switch in the "Short" position, all three LEDs—"Y-G," "G-R," and "R-B"—should be "on" to indicate no shorts or crosses between conductors. If any of the three LEDs are "off," there is a short or cross between the indicated conductors.

   Plug the other end of the modular cord into jack J2. With the battery switch in the "Cont" position, all four LEDs—"Y," "G," "R," and "B"—should be "on" to indicate continuity in the indicated conductors. If any of the four LEDs are "off," there is a lack of continuity in the indicated conductor(s).
2. *Station Wiring "Short" Test*

   Plug one end of any modular cord into jack J1. Plug the other end of the cord into a modular wall jack of "dry" station wiring. With the battery switch in the "Short" position, all three LEDs—"Y-G," "G-R," and "R-B"—should be "on" to indicate no shorts or crosses. If any of the three LEDs are "off," there is a short or cross between the indicated conductors.
3. *Station Wiring "Ground" Test*

   Plug one end of any modular cord into jack J2. Plug the other end of the cord into a modular wall jack of "dry" station wiring. Attach the "Ground" alligator clip to station ground. With the battery switch in the "Cont" position, the "Y," "G," "R," and "B" LEDs should be "off" to indicate that the station wiring conductors—yellow, green, red, and black, respectively—are not grounded. If any of the four LEDs are "on," there is a ground in the indicated station wiring conductor(s). Note that it may be desirable to have one or more of the conductors grounded; hence, the test set can be used to verify the ground.
4. *Station Wiring Continuity, Voltage, Current, and Polarity Test*

   Plug one end of any modular cord into jack J2. Plug the other end of the cord into a modular wall jack of "wet" station wiring. With the battery switch in the "Off" position, the "R" LED will always be "off" for the following tests, if the wiring tests good. The "Y," "G," and "B" LEDs "on" indicate continuity, correct tip and ring voltage polarity, voltage available for talk battery between the red and green conductors, and voltage available for lamps between the yellow and black conductors. The "Y" LED "off" and the "B" LED "on" indicates a dc voltage between the yellow and black conductors, with positive on the yellow and negative on the black. The "B" LED "off" and "Y" LED "on" indicates a dc voltage between the yellow and black conductors, with positive on the black and negative on the yellow. Both "Y" and "B" LEDs "off" indicates insufficient voltage between the yellow and black conductors, or lack of continuity in either the yellow or the black conductor. The "G" LED "off" indicates tip and ring reversal, lack of continuity in the green and red conductors, or insufficent voltage available between the red and green conductors.

Optional test cords may be used as "go/no-go" indicators of voltage levels, hence verifying not only the presence or absence of voltage, but also giving an indication of sufficient voltage. By using the cord labeled "HSE" (house) in place of the modular test cord mentioned in the beginning of this section, the "Y," "G," and "R" LEDs "on" indicates greater than 46 volts dc and 20 mA current available across the red and green conductors for talk battery; and greater than 6 volts dc and 20 mA current available across the yellow and black conductors for lamps, such as night lights, etc.

By using the cord labeled "KS" (key system) in place of the modular test cord mentioned in the beginning of this section, the "Y," "G," and "R" LEDs "on" indicates greater than 22 volts and 20 mA current available across the red and green conductors for talk battery; and greater than 10 volts and 20 mA current available across the yellow and black conductors for lamps. All other "status" indications from the LEDs remain the same as when using the standard modular test cord.

5. *Test Set Verification*

With the battery switch in the "Short" position, and with no cords in jacks J1 or J2, the "Y-G," "G-R," and "R-B" LEDs should be "on" to indicate a good battery. If all LEDs are "off," a bad battery or defective test set is indicated.

APPENDIX **A**

Troubleshooting

This appendix contains a list of common troubles you may encounter during the life of your telephone. Remember, if you do not own the telephone equipment, you are not allowed to repair it. Call your local telephone company and their repair service personnel will repair it without charge. Quite a number of the listed troubles stem directly from their experiences. Even if you do not own your telephone equipment, this troubleshooting chart may enable you to report your trouble to the repair service more accurately and enable them to repair it more quickly.

Keep in mind that we are not trying to make an electrical engineer out of you. With this chart, however, you should be able to either fix your set or determine if it has to go to a repair shop. In the beginning of this book, you were told that it could save you money. This is where you will see how.

Trouble	Possible Cause	Remedy
No dial tone.	Telephone set not plugged in.	Check wall receptacle. Check set receptacle.
	Handset not plugged in.	Check handset receptacle.
	Defective wall cord.	Replace wall cord.
	Defective handset cord.	Replace handset cord.
	Defective receiver.	Replace receiver in handset.
	Defective cradle switch.	Check for stuck plungers when handset is off the cradle.

Trouble	Possible Cause	Remedy
No dial tone (cont).	Insufficient current.	Check current supplied by central office. It should be greater than 21 milliamperes. If the current is low, call the telephone company.
Dial tone does not stop when dialing.	Tip and ring conductors reversed.	Reverse tip and ring conductors. (This is unique to DTMF telephone sets.)
	Defective dial pulse contacts.	Replace dial.
	Dial pulsing off speed.	Replace dial. (There is an adjustable speed-controlling governor, but it requires special tools to adjust.)
	Insufficient current.	Check current supplied by central office. It should be greater than 21 milliamperes. If the current is low, call the telephone company.
No DTMF tones.	Defective DTMF dial.	Replace DTMF dial.
	Tip and ring conductors reversed.	Reverse tip and ring conductors.
	Defective polarity guard.	This is inside the telephone set. If dial tone is present, reverse the tip and ring conductors. If the dial tone disappears, you may have a defective polarity guard. Replace polarity guard, if possible, or replace telephone set.
Receiver "clicks" when dialing.	Defective dial off-normal contacts.	Replace dial.
Bell taps ("dings") when rotary dialing.	Tip and ring conductors reversed.	Reverse tip and ring conductors.
	Defective multiparty ringer.	Check for improperly connected ringer, wrong frequency ringer, incorrect ringer spring bias, or replace ringer.

Trouble	Possible Cause	Remedy
Rotary dialing produces wrong numbers.	Dial pulsing off speed.	Replace dial. (Note: Manually forcing dial to return too fast can also cause this trouble, as will not allowing sufficient interdigital time. Hesitate between dialing digits and allow dial to return normally.
	Insufficient current.	Check current supplied by central office. It should be greater than 21 milliamperes. If the current is low, call the telephone company.
Telephone will not ring but has audible dial tone when receiver is off hook.	Volume control off.	Check volume control.
	Defective multiparty ringer.	Check for improperly connected ringer, wrong frequency ringer, incorrect ringer spring bias, or replace ringer. Also check for good ground from telephone set to earth.
	Defective ringer.	Check for bent clapper, check each ringer coil for continuity, check for loose connections, or replace ringer.
	Defective ringer capacitor.	Replace ringer capacitor.
	Defective network.	Replace network.
	Defective cradle switch.	Check for contacts not closing when handset is in cradle. Adjust contacts or replace switch.
Telephone rings on wrong line (multi-party line only).	Defective multiparty ringer.	Check for improperly connected ringer, wrong frequency ringer, incorrect ringer spring bias, or wrong value ringer capacitor. Exchange set for one with correct ringer equivalence.
Telephone rings but ringer volume is low.	Volume control off.	Check volume control.
	Clapper obstructed.	Check clapper and remove any obstruction.

Trouble	Possible Cause	Remedy
Telephone rings but ringer volume is low (cont).	Telephone resting on sound-absorbing material such as soft pillow, etc.	Place telephone on hard surface such as table, etc.
	Too many extension sets on your telephone line.	Check total ringer equivalence number; must not exceed 5.
Telephone rings, you can hear caller but caller cannot hear you.	Defective transmitter.	Replace transmitter.
	Defective handset cord.	Replace handset cord.
	Defective dial contacts.	Off-normal dial contacts shunting transmitter with dial at rest. Replace dial.
	Defective cradle switch.	Contacts do not open when handset is raised. Adjust or replace cradle switch.
Telephone rings but you cannot hear caller and do not know if caller can hear you.	Defective receiver.	Replace receiver.
	Defective handset cord.	Replace handset cord.
	Defective dial contacts.	Off-normal contacts are open in receiver circuit with dial at rest. Adjust contacts or replace dial.
	Defective cradle switch.	Contacts do not open when handset is raised. Adjust or replace cradle switch.
	Defective network.	Replace network.
Other party says your voice is distorted.	Defective transmitter.	Gently tap transmitter end of handset on hard surface several times. (This trouble is common with candlestick-type telephones.) Replace transmitter.
	Bad line.	Call the telephone company. If you are on long distance, inform the operator.
	Loose connections.	Check connections in your handset and telephone.

TROUBLE—GENERAL

If your set is in warranty, return it for repair. You may invalidate your warranty if you try to repair it yourself.

When repairing your set, if you can obtain an identical set that is working, you can substitute parts from the good set until you determine the trouble in the defective set. Unless you have good skills, adjustments may prove impossible. Replace components instead of adjusting. If you doubt your ability, or lack test equipment and tools, take your set to a repair shop and have it repaired correctly.

Keep in mind that you are *not* allowed to do any repair work on equipment owned by the telephone company.

APPENDIX **B**

Numerical List of Area Codes

For your convenience, a list of area codes in numerical sequence is included. When dialing long distance, save the operator a bit of trouble and dial "1 + area code + telephone number."

AREA CODE	LOCATION	AREA CODE	LOCATION
201	New Jersey	305	Florida
202	District of Columbia	306	Saskatchewan
203	Connecticut	307	Wyoming
204	Manitoba	308	Nebraska
205	Alabama	309	Illinois
206	Washington	312	Illinois
207	Maine	313	Michigan
208	Idaho	314	Missouri
209	California	315	New York
212	New York	316	Kansas
213	California	317	Indiana
214	Texas	318	Louisiana
215	Pennsylvania	319	Iowa
216	Ohio		
217	Illinois	401	Rhode Island
218	Minnesota	402	Nebraska
219	Indiana	403	Alberta
		403	Northwest Territories (Dist. of Mackenzie) (Temp.)
301	Maryland	403	Yukon (Temp.)
302	Delaware	404	Georgia
303	Colorado	405	Oklahoma
304	West Virginia		

AREA CODE	LOCATION	AREA CODE	LOCATION
406	Montana	701	North Dakota
408	California	702	Nevada
412	Pennsylvania	703	Virginia
413	Massachusetts	704	North Carolina
414	Wisconsin	705	Ontario
415	California	707	California
416	Ontario	709	Newfoundland
417	Missouri	712	Iowa
418	Quebec	713	Texas
419	Ohio	714	California
		715	Wisconsin
501	Arkansas	716	New York
502	Kentucky	717	Pennsylvania
503	Oregon		
504	Louisiana	801	Utah
505	New Mexico	802	Vermont
506	New Brunswick	803	South Carolina
507	Minnesota	804	Virginia
509	Washington	805	California
512	Texas	806	Texas
513	Ohio	807	Ontario
514	Quebec	808	Hawaii
515	Iowa	809	Caribbean Islands
516	New York	812	Indiana
517	Michigan	813	Florida
518	New York	814	Pennsylvania
519	Ontario	815	Illinois
		816	Missouri
601	Mississippi	817	Texas
602	Arizona	819	Quebec
603	New Hampshire		
604	British Columbia	901	Tennessee
605	South Dakota	902	Nova Scotia
606	Kentucky	902	Prince Edward Island
607	New York	903	Mexico
608	Wisconsin	904	Florida
609	New Jersey	906	Michigan
612	Minnesota	907	Alaska
613	Ontario	912	Georgia
614	Ohio	913	Kansas
615	Tennessee	914	New York
616	Michigan	915	Texas
617	Massachusetts	916	California
618	Illinois	918	Oklahoma
		919	North Carolina

APPENDIX **C**

Federal Communications Commission Rules

When you purchase any items that you intend to connect to the telephone line, you must be sure you are not violating any FCC rules. In the words of the FCC issued in a Public Notice dated May 24, 1979:

PUBLIC NOTICE

Federal Communications Commission ■ 1919 M Street, NW. ■ Washington, D.C. 20554

For recorded listing of releases and texts call 632-0002

For general information call 632-7260

17840

May 24, 1979 - CC

GRANDFATHERING PERIOD ENDING FOR TELEPHONES, DATA AND ANCILLARY DEVICES

Manufacturers, distributors and retailers of telephones, data and ancillary devices are reminded that the grandfathering period for these devices expires on July 1, 1979. Such devices installed after that date must bear FCC registration labels.

The purpose of the grandfathering period was to give manufacturers a reasonable period of time to obtain FCC registration and to work off inventories of grandfathered items. A list of grandfathered devices has been published by the FCC. Manufacturers and distributors of such devices who wish to continue to market them and have not obtained registration may register items appearing on the Grandfathered Telephone Equipment List by filing a short-form application. A registration kit consisting of FCC Form 730, Instructions and Part 68 of the Commission's Rules may be obtained from:

Federal Communications Commission
Domestic Branch, Room A-304
1919 M Street, N.W.
Washington, D.C. 20554

Telephone: (202) 632-6440

- FCC -

What is "grandfathered" telephone equipment? "All terminal equipment [other than pbx and key telephone systems, both of which we are not concerned with in this book] and protective circuitry [again we are not concerned with this type of equipment in this book] of a type directly connected to the telephone network as of October 17, 1977, may be connected thereafter up to July 1, 1979, and may remain connected for life, without registration, unless subsequently modified." These words, contained in Part 68 of the FCC Rules and Regulations, also tell us that any telephone equipment you purchase must be registered in order to be legally connected to the telephone network after July 1, 1979. If you have an old piece of telephone equipment and want to determine if it has been grandfathered, you can obtain a copy of the FCC Grandfathered Telephone Terminal Equipment List for a duplication charge from:

Down Town Copy Center
1114 21st Street NW
Washington, DC 20036
Telephone: 1-202-452-1422

Or, you can call the FCC at 1-202-632-6440 to determine if your telephone is on the list.

FCC REGISTERED TELEPHONE EQUIPMENT

To connect FCC registered telephone equipment, all that is required is to inform your telephone company of the make, model number, FCC registration number, and ringer equivalence number (REN), all of which will be found on a label attached to the outside of the equipment.

CONNECTION OF CUSTOMER-OWNED TELEPHONE EQUIPMENT

On July 1, 1979, all telephone companies began installing the new modular (miniature) receptacles as the nationwide standard for the connection of telephones. Most of the grandfathered telephones are equipped with large four-prong plugs. If your home receptacle is the new minimodular type and you want to connect a telephone with a large four-prong plug, then an adapter will be required. Conversely, if your telephone is the registered type with the minimodular plug and your home receptacle is the large four-prong type, another type of adapter will be required. Either type of adapter should be available from electronic specialty stores carrying telephone equipment.

INSTALLATION OF RECEPTACLES

Unless specifically permitted by local tariffs, all receptacles must be installed by the telephone company.

As of the date this book was written, only the Rochester Telephone Company in New York will allow you to install your own receptacles. Many other telephone companies are contemplating such action. Check with your local telephone company to determine its rules and regulations.

On a nationwide basis, you can install an FCC registered protective coupler at the point the telephone company facilities enter your home or business. You can then install your own premise wiring. Again, please check with your telephone company to determine local rules and regulations.

If you have the telephone company install a receptacle, there will usually be a one-time service charge. Try to plan your entire home or business installation so only one visit by the telephone company will be necessary. This will result in the minimum possible cost to you, particularly if a number of receptacles are desired.

TELEPHONE COMPANY DISCOUNTS FOR CUSTOMER-OWNED EQUIPMENT

When the U. S. Supreme Court, in October 1977, permitted the FCC Telephone Terminal Equipment Registration Program to go into effect, the telephone companies filed tariffs with each state public utility commission permitting a discount ranging from about 50 cents to one dollar per month off of the normal extension rates for each customer-owned telephone. All such charges are approved by the various state public utility commissions, not the FCC.

CUSTOMER-OWNED EQUIPMENT ON MULTIPARTY LINES

The FCC found that the technical characteristics of party-line services vary so much that it was impossible to develop nationwide standards. Using devices designed for single-party service on multiparty lines can cause many problems; for example, an answering machine connected to a multiparty line most likely will answer all calls coming in on the line. Telephones not specifically designed for a given party line may not ring and may improperly cause the central office to bill another party, or they may cause severe service impairments. Consequently, the FCC decided that only telephone companies may provide equipment for direct connection to multiparty lines. The customer cannot provide his own equipment if he has multiparty service.

REPAIR OF TELEPHONE EQUIPMENT

The FCC has set down rules in *"Part 68—Connection of Terminal Equipment to the Telephone Network"*

> Section 68.216 *Repair of Registered Terminal Equipment and Registered Protective Circuitry.*
>
> Repair of registered terminal equipment and registered protective circuitry shall be accomplished only by the manufacturer or assembler thereof or by their authorized agent; however, routine repairs may be performed by a user, in accordance with the instruction manual if the applicant certifies that such routine repairs will not result in noncompliance with the rules and regulations in Subpart D of this Part.

The "applicant" referred to in the preceding paragraph is the company that applied for registration of the equipment. "Subpart D" referred to is the section of Part 68 concerned with "Conditions for Registration."

In a few words, do not attempt to repair telephone equipment that you do not own; and, if you own the equipment, only repair it if the manufacturer's instructions tell you how and to what extent repairs are permitted.

APPENDIX **D**

Index of Manufacturers and Suppliers

The following list is not all-inclusive. There are hundreds of companies that will provide telephone equipment for you. Perhaps the list will be of some help to you if no local sources can be found. Some of the listed companies will not sell directly to you, but instead will refer you to the retailer nearest you.

A

Action Systems Co.
34 Cambridge Street
Meriden, CT 06450

Aerotron, Inc.
P.O. Box 27500
Raleigh, NC 27611

Allied Communications Co.
18 East 41st Street
New York, NY 10017

Allied Telecommunications Equipment Co.
206 West 40th Street
New York, NY 10018

American Telephone & Telegraph Co.
195 Broadway
New York, NY 10007

Anaconda Co., Wire and Cable Div.
Greenwich Office Park 3
Greenwich, CT 06830

Arrow Fastener Co., Inc.
271 Mayhill Street
Saddle Brook, NJ 07662

B

BBL Industries
2830 Clearview Place
Atlanta, GA 30340

Bramco Controls
1711 Commerce Drive
P.O. Box 706
Piqua, OH 45356

Buckeye Telephone/Supply Co.
1250 Kinnear Road
Columbus, OH 43221

Buscom Systems, Inc.
1230 Mt. View Aluiso Road
Sunnyvale, CA 94086

C

Chrome Phone Products
6080 Jericho Turnpike
Commack, NY 11725

Comex Systems
P.O. Box 278
Hudson, NH 03051

Communications Products, Inc.
1401-B South Floyd Road
Richardson, TX 75080

Communications Specialists
11409 Chandler Boulevard
North Hollywood, CA 91601

Cordura Marketing, Inc.
10 South LaSalle Street
Chicago, IL 60603

Coreco Research Corp.
370 Seventh Avenue, Suite 301
New York, NY 10001

D

Data-Phone Corp.
2590 East Devon Avenue
Des Plaines, IL 60018

Data Signal
2403 Commerce Land
Albany, GA 31707

Datotek, Inc.
13740 Midway Road
Dallas, TX 75240

Davis-Denver, Inc.
2600 South Parker Road
Building T
Denver, CO 80232

Digital Telephone Systems
P.O. Box 4222
San Rafael, CA 94903

F

Fan-Tron Corp.
4023 West 30th
Little Rock, AR 72204

Floyd Bell Associates, Inc.
1333 Chesapeake Avenue
Columbus, OH 43212

Fold-A-Phone, Inc.
3180 Expressway Drive, South
Central Islip, NY 11722

Ford Industries
5001 Southeast Johnson Creek
Boulevard
Portland, OR 97206

Fracom Enterprises
2130 West Clybourn
Milwaukee, WI 53233

G

Gimix, Inc.
1337 West 37th Place
Chicago, IL 60609

Glenayre Electronics, Inc.
8 12th Street
P.O. Drawer L
Blaine, WA 98230

Goldberg Co.
144-19 75th Road
Flushing, NY 11367

GTE Automatic Electric, Inc.
Communications Systems Div.
400 North Wolf Road
P.O. Box 2059
Northlake, IL 60164

Gutzmer International
7215 Convoy Court
San Diego, CA 92111

I

ICP Marketing
3031 Tisch Way, Suite 750
San Jose, CA 95128

Integrated Circuits Packaging, Inc.
3008 Scott Boulevard
Santa Clara, CA 95050

Integrated Systems Technology
P.O. Box 2585
Garland, TX 75041

ITT Consumer Specialty Products
133 Terminal Avenue
Clark, NJ 07066

L

Lear Siegler, Inc.
Electronic Instrumentation Div.
714 North Brookhurst Street
Anaheim, CA 92803

Livermore Data Systems, Inc.
2050 Research Drive
Livermore, CA 94550

M

Magnetic Electronics
211 Orr Avenue
Opelika, AL 36801

Maxxima Electronics Corp.
21736 Roscoe Boulevard
Canoga Park, CA 91304

Microelectronic Systems Corp.
29245 Stevenson Highway
Madison Heights, MI 48071

Motorola, Inc.
1303 East Algonquin Road
Schaumburg, IL 60196

N

National Telephone & Electronics, Inc.
7561 NW 77th Terrace,
Miami, FL 33166

Northern Telecom
640 Massman Drive
Nashville, TN 37201

North Supply Co.
10951 Lakeview Avenue
Lenexa, KS 66219

O

Omicron Data Systems
4480 Cote De Liesse Road
Montreal, Quebec, Canada

Onyx Telephone Center
12 West 32nd Street
New York, NY 10001

P

Panasonic Co.
One Panasonic Way
P.O. Box 1503
Secaucus, NJ 07094

Phone Control Systems
92 Marcus Avenue
New Hyde Park, NY 11040

Phone Devices Corp.
524 South Michigan Avenue
Chicago, IL 60605

Phone-Mate
325 Maple Avenue
Torrance, CA 90503

Phonetele, Inc.
7900 North Sepulveda Boulevard
Van Nuys, CA 91405

Q

Quasar Microsystems, Inc.
448 Suffolk Avenue
Brentwood, NY 11356

R

Radio Shack/Div. Tandy Corp.
1600 One Tandy Center
Fort Worth, TX 76102

Reach Electronics, Inc.
P.O. Box 308
Lexington, NE 68850

Record-A-Call
16250 Gundry Avenue
Paramount, CA 90723

Recoton Corp.
46-23 Crane Street
Long Island City, NY 11101

Regency Communications, Inc.
1227 South Patrick Drive
Satellite Beach, FL 32937

Remco International
1723 West Howard Street
Evanston, IL 60202

Robins Industries Corp.
75 Austin Boulevard
Commack, NY 11725

Royce Electronics Corp.
1747 Levee Road
North Kansas City, MO 64116

S

Sanyo Electric, Inc./Electronics Div.
1200 West Artesia Boulevard
Compton, CA 90220

Saxton Products
215 North Route 303
Congers, NY 10920

Seaboard Electronics Co.
70 Church Street
New Rochelle, NY 10805

Secode Electronics
908 Dragon Street
Dallas, TX 75207

Smith-Gates Corp.
1451 New Britain Avenue
Farmington, CT 06032

Specialized Systems, Inc.
11558 Sorrento Valley Road
Building 7
San Diego, CA 92121

Speedcall Corp.
2020 National Avenue
Hayward, CA 94545

Standard Communications Corp.
108 West Victoria
P.O. Box 92151
Los Angeles, CA 90009

T

Technical Communications Corp.
56 Winthrop Street
P.O. Box 1070
Concord, MA 01742

Technology Applications Corp.
560 San Antonio Road
Spectra Suite 204
Palo Alto, CA 94306

Telcer Telephone USA
319 Lynn Way
Lynn, MA 01901

Telco Products Corp.
44 Sea Cliff Avenue
Glen Cove, NY 11542

Telecomputer Corp.
211 Orr Avenue
Opelika, AL 36801

Teleconcepts, Inc.
22 Culbro Drive
West Hartford, CT 06110

Tele Devices Corp.
17 Clinton Street
Plattsburgh, NY 12901

Telephonic Equipment
17281 Eastman Street
Irvine, CA 92660

Tele-Tender Systems Corp.
9 East 37th Street
New York, NY 10016

Tel Products, Inc.
7801 Bass Lake Road
Minneapolis, MN 55428

Tinekit
23715 Mercantile Road
Cleveland, OH 44122

Travel Electronics International
17921 Skypark Boulevard
Irvine, CA 92714

Tridar
P.O. Box 61929
Sunnyvale, CA 94086

TRS Marketing, Inc.
137 East Savarona Way
Carson, CA 90746

U

Universal Phone Corp.
1746 Levee Road
North Kansas City, MO 64116

Utility Verification Corp.
66 Commack Road
Commack, NY 11725

Z

Zoom Telephonics
65 Franklin Street
Boston, MA 02110